Handbook of Biofertilizers and Microbial Pesticides

Handbook of Biofertilizers and Microbial Pesticides

DR. M.S. VORA
Research Scientist (Microbiology)
Biofertilizer Project & Head

MRS. H.N. SHELAT
Assistant Research Scientist (Microbiology)

DR. R.V. VYAS
Associate Professor (Microbiology)
& PI, EPN Project, Invertebrate Pathologist

Department of Microbiology, B. A. College of Agriculture
Anand Agricultural University, Anand - 388110, Gujarat, India

SSPH

SATISH SERIAL PUBLISHING HOUSE
403, Express Tower, Commercial Complex
Azadpur, Delhi-110033 (India)
Phone : 011-27672852, Fax : 91-11-27672046
e-mail : info@satishserial.com, hkjain1975@yahoo.com
Website : www.satishserial.com

Published by :

SATISH SERIAL PUBLISHING HOUSE
403, Express Tower, Commercial Complex, Azadpur, Delhi-110033 (INDIA)
Phone : 011-27672852 Fax : 91-11-27672046
E-mail : info@satishserial.com, hkjain1975@yahoo.com

Reprint, 2021

ISBN : 978-81-89304-41-6

Composed, Designed & Printed in India

ABOUT THE AUTHORS

Dr. Mayank S. Vora, Ph.D. working as a Research Scientist at Department of Microbiology, Anand Agricultural University, Anand (Gujarat) India, has about 33 years experience of research and teaching in the field of microbiology.

He has published 33 research papers in National and International journals and produced 35 recommendations of various biofertilizers for the benefit of Gujarat Farmers. He is a recipient of prestigious Sardar Patel, Award for his contribution in the field of PSM an also been recipient of Bharat Jyoti Award. Dr. Vora has carried out 5 research projects on bifertilizers. He is also a resource person for imparting training on biofertilizers in various state training programmes. He has undertaken most advanced training in Legume/Rhizobium Technology from NiFTAL, Hawaii, USA. He is also acting as external examiner at postgraduate level in many traditional and state agricultural universities.

Mrs. Harsha Shelat, M.Sc. working as a Assistant Research Scientist at Department of Microbiology, Anand Agricultural University, Anand (Gujarat) India, has about 23 years experience of research and teaching in the field of microbiology. She has published more than 35 research papers in National and International journals and produced 35 recommendations of various biofertilizers for the benefit of Gujarat Farmers. She has designed and formulated few booklets as well as leaflets on biofertilizers for popularizing technology. She is a recipient of prestigious Sardar Patel, Award for his contribution in the field of PSM. She is also a resource person for imparting training on biofertilizers in various state training programmes.

Dr. Rajababu V. Vyas, Ph.D., working as Associate Professor (Microbiology) & PI, EPN Project at B. A. College of Agriculture, Anand Agricultural University, Anand (G.S.), India and has 19 years research experience on Microbial control of insects and phyto-nematodes with regards to isolation, development of mass production technique, laboratory and field-testing of invertebrate pathogens. He has published more then 85 research papers in national and international journals, 4 book chapters and review articles in international reputed, edited publications. He has also produced many recommendations, 7 for scientific community and 6 for farming community to provide non-chemical management practices for agricultural pest management. He has received 2 awards, HariOm Ashram Prerit Best Research Paper for Teachers and Nematological Society of India Best presentation. He has guided 9 postgraduate students so far. Dr. Vyas has received funds from ICAR and GSFC and carried out 3 research projects on biological control and presently has few projects on hand. He is also a resource person for imparting training on microbial pesticides in various state and national training programmes. He is also Advisor Third World Academy of Sciences (TWAS), Trieste, Italy; ICAR, New Delhi; Japan Science & Technology Corporation (JST), Kawaguci-City, Japan for project and report evaluation. He is member of seven scientific organizations in India and extending his appropriate services. He is also acting as external examiner at postgraduate level in many traditional and state agricultural universities.

FOREWORD

Since the ancient time, mankind has taken advantage of microorganisms appreciating their beneficial role in living systems. They have great impact on man, animal and plant life. With the growing appreciation in the later half of the last century, their role in agriculture was recognized as biofertilizer and subsequently as microbial pesticides. The study on microbes due to commercialization of these two beneficial microbes in world market has taken a new turn for the high-tech researches in today's agriculture. These bio inoculants are easy for augmentation due to simplicity in production technology developed in recent years. They are natural allies and effective for integrated nutrient and pest management systems with high degree of safety to higher animals and plant life. Moreover, they are indigenous, environmentally compatible and safe for better crop productivity and protection.

I hope the publication on ***Handbook of Biofertilizers and Microbial pesticides*** by microbiologists, Dr M.S. Vora, Mrs Harsha Shelat and Dr R. V. Vyas of B. A. College of Agriculture, AAU, will help students and researchers alike. They have shared their knowledge and experiences by compiling and publishing this compendium. I'm sure this publication would be highly useful to research workers of public and private enterprises as well as undergraduate and post graduate students as a source of valuable and new informations on microorganisms for encouraging agricultural research and developments.

Anand Agricultural University
Anand - 388 110
January, 2007

(M.C. VARSHNEYA)
Vice Chancellor

PREFACE

The world population is likely to double during next 50 years, so also the demand of more food. At present, India produces about 17 million tonnes of food grains. Indiscriminate use of different agrochemicals including chemical fertilizers has become an integral part of today's agriculture. To meet the food demand of ever increasing population, the requirement of only nitrogenous fertilizers is estimated to be about 11.4 million tonnes. Chemical fertilizer is not only costly but growing awareness of environment pollution and limitations of nonrenewable resources may introduce additional constraints. With almost twice the quantity of plant nutrients being removed from soil than added through fertilizers, the growing nutrient imbalance poses a major threat to sustain soil health and crop productivity. To minimize the use of these inputs without affecting the overall production and the ecosystem, it is necessary to popularize the benefits of ecofriendly microbial inoculants, in our Agriculture. Indian soils are increasingly becoming poorer in organic matter and there is growing awareness about regular application of organic manures including FYM, compost, deoiled cakes, green manures etc to sustain soil productivity. Organic manures can supplement plant nutrient supplies but are required in large quantity, which is difficult for farmers to procure particularly under condition of unorganized collection, poor conservation methods and of poor nutritional quality. Incorporation of biofertilizers and microbial pesticides in combination with organic manures and inorganic fertilizers can completely obviate the detrimental effects of current agricultural practices.

This book covers major biofertlizers including ***Rhizobium, Azotobacter, Azospirillum, Acetobacter,*** **Phosphate Solubilizing microorganisms** and major microbial pesticides viz; ***Trichoderma, Paeciliomyces,*** **Entomopathogenic nematodes, NPV and *Bacillus thuringiensis.*** The book also includes equipments and raw material, production technology, project profile, standards and quality control, application technology etc.

The book has been specially prepared with more emphasis on providing hands in basic microbiological techniques with information on useful microorganisms and with detailed information required for setting up a pilot scale production facility together with preparation and down steam processing of nitrogenous biofertilizers, phosphatic biofertilizers and different microbial pesticides. We hope this will be extremely useful to the new entrepreneurs, teachers, scientists, researches, undergraduate and postgraduate students.

We gracefully acknowledge the support and valuable suggestions of Professor M. C. Varshneya who was kind enough to permit us to publish this book. The authors express their sincere thanks to Dr. A.R. Pathak and Dr. A.M.Shekh for their encouragement and support. The authors are indebted to staff and fellows of department for untiring effort in bringing out this volume.

Anand Agricultural University
Anand - 388 110
January, 2007

M. S. Vora
H. N. Shelat
R. V. Vyas

"A starving man thinks first of satisfying his hunger before anything else. He will sell his liberty and all for the sake of getting a morsel of food. Such is the position of millions of people of the India. For them, liberty, God and all such words are merely letters put together without slightest meaning."

" Mahatma Gandhi"

CONTENTS

Chapter 1

Biofertilizers Marketing and Prospects for Commercialization

INRODUCTION

Indian agriculture has shifted from traditional to intensive farming with indiscriminate use of chemical fertilizers and pesticides, which in turn has made our soils largely non-productive and contaminated the ground water. Indian soils are also becoming poor in organic matter and there is a growing concern for continuing regular application of organic manure and recycling crop residues to sustain productivity and high responses to NPK fertilizer. Forest areas have been adversely affected with large-scale deforestation activities, which have considerably reduced the forest cover. The impending energy crisis due to fast depleting mineral oil reserves has necessitated the search for renewable energy resources to ease the pressure on fuel energy. Chemical fertilizers always falling short by 8-10 million tons are staggering, considering country's future demand for food. Incorporation of biofertilizers or microbial inoculants in combination with organic manure can reduce the detrimental effects of the current agricultural practices in an eco-friendly manner. Biofertilizers are best described as microorganisms, which add, conserve and mobilize the crop nutrients in the soil. Use of biofertilizers such as *Rhizobium*, *Azospirillum*, *Azotobacter*, blue green algae, *Azolla*, mycorrhizae and phosphate solubilizers can facilitate management of crop nutrients leading to long term sustainability in crop production. The most striking relationship that these have with plants is symbiosis, in which the partners derive benefits from each other. Their addition also increases the fertilizer use efficiency of the crop plants. With ever increasing population, it is necessary to increase crop productivity per unit area without causing further deterioration in soil health and maintaining micronutrient balance.

Later half of the last century has witnessed remarkable development in the field of biofertilizers. The efforts have created a *cafeteria* of biofertilizers and revealed variety of attributes of these microbes implicating them in nitrogen fixation, phosphate solubilization, addition of organic matter, secretion of growth factors, improving the physical and chemical properties of the soil, better utilization of the chemical fertilizers, bioconcentration of nutrients in the rhizosphere and acceleration of the process of composting.

Table 1.1. Microorganisms having promise to boost agricultural production

Function	Organism	Crop
Nitrogen Fixation	*Rhizobium meliloti, R.leguminosarum,*	Arhar, Pea, Lentil Gram, Green Gram
	R.irifolii, R.lupini, R.japonicum,	Rajmah, Cowpea, Berseen, Leucerne, Groundnut, Soybean
	Azotobacter choroococcum, A.vinelandii,	Wheat, Barley
	A.beijjerinkii,	Maize, Cotton, Paddy
	Acetobacter diazotrophicus	Sugarcane
	Azospirillum lipoferum, A.amazonense,	Maize, Sugarcane
	A.seropedica, A.americanum,	Sorghum, Paddy, Pearl millet
	Cyanobacteria (blue green algae) Tolypothrix, Anabaena azollae	Transplanted Paddy
Phosphate Solubilization	*Bacillus polymyxa, B.megaterium, B. coagulans*	All crops

Contd...

Table 1.1. Contd...

	Pseudomonas striata, Aspergillus awamori, *Penicillum digitatum*	
Nutrient Translocation	Endomycorrhizae: *Glomus, Gigaspora, Accaulospora*	Forest tree species, all crops
	Ectomycorrhizae: *Laccaria, Pisolithus*	Forest tree species
Organic Matter Decomposition	*Cellulolytic:* Fungi: *Trichoderma, Chaetomium Aspergillus* Bacteria: *Bacillus, Cellulomonas, Cytophaga*	
	Lignolytic: Fugni: *Clavaria, Cephalosporium*	
Growth Accelerators	Plant growth promoting Rhizobacteria (PGPR), *Pseudomonas, Xanthomonas, Flavobacterium*	All Crops

More than twenty-five countries including both developed countries like USA, UK and China and other developing countries like Thailand, Uganda and Zimbabwe have already initiated production and utilization of biofertilizers on a large scale.

Several forms of microbial inoculants viz. agar based; broth cultures; frozen concentrate; carrier based; granular soil inoculum; porous gypsum granules; peat granules and freeze dried inoculants can be used for inoculating seed or soil. The choice of type of inoculum depends upon demand, availability of carrier material, survival period and organizational setup.

Biofertilizers are applied at seed or soil dressing and through broadcasting. The diazotrophic biofertilizers provide a proxy to nitrogenous fertilizers up to 35 kg N ha^{-1}. They also exert supplementation effect, which is more pronounced at lower fertilization levels. Combined application of compatible biofertilizers substantially increases the agronomic benefits.

MARKET

Most of the production of biofertilizers in the country is being done in the public sector by research institutions, universities and National Biofertilizer Development Centre. Few state and co-operative fertilizer units also have ventured into field. Surprisingly, the involvement of private sector is extremely limited in spite of it being a low investment and high benefit technology. Almost all the production units in the country are producing only the bacterial biofertilizers and the present day annual production is estimated to be around 9,.000 metric tones, which is far below the demand.

On the other hand our domestic production of nitrogen fertilizers is only around 8.6 million tons with a infrastructure investment of Rs. 1,20,000 million. This, together with an import of 2.34 million tons and a total subsidy of Rs.48, 000 million is able to meet only about 2/3 of the total fertilizer requirement of the country. The consumption of fertilizers is also not uniform and seems to be directly proportional to the size of the land holdings. About 68 million farmers owning 65 per cent of the total arable land, use only 30 per cent of the total fertilizers consumed in the country. It is this group which can be easily motivated to adopt biofertilizer technology. The fertilizer (NPK) production and requirement, in million metric tones of our country is given in Table 1.2.

Table 1.2. Fertilizer requirement and production in India Projections (Million tonnes)

Year	2000	2011	2031	2051
Requirement	19.0	22.2	27.3	31.3
Production	14.9	15.8	20.9	23.9
Gap	4.1	6.4	6.4	7.2

Depending upon the total area under different crops amenable to biofertilizer inoculation, the total annual requirement of the bacterial biofertilizers that can be generated within 5 years of rigorous campaigning comes to about 17,000 tons.

On a conservative estimate, even a 10 per cent saving through the use of biofertilizers is expected to result in an annual saving of 1.094 million tons of nitrogenous fertilizers costing around Rs.5, 500 million.

TECHNOLOGY

Unlike chemical fertilizers, all biofertilizers contain viable population of selected microbes, which are able to colonize the soil ecosystem. Their production on large scale needs favourable growth conditions and defined nutrient media. Bacteria and fungi being heterotrophs, need organic substrates, whereas, algae are photolithotrophs and require inorganic nutrients and light.

Table 1.3. Demand of biofertilizers expected to be generated in five years:

Biofertilizer	Packets (x 10^6)	Weight (tons)
Rhizobium	21.0	4,200
Azotobacter	39.0	7,800
Others	26.0	5,200
Algal	24.0	12,000
Total	110.0	29,200

NOTE: Weight of each packet sufficient for 1 acre: bacterial, 200g: algal 500 g.

Table 1.4. Area, nitrogenous fertilizers consumed and saving through biofertilizer application in different crops.

Crop	Area of cultivation (ha)	Nitrogen fertilizer used (tons)	Fertilizer saving* (tons)
Paddy	42.59	4.30	0.430
Wheat	25.38	3.02	0.302
Sugarcane	3.82	0.92	0.092
Cotton	8.30	0.62	0.062
Other crops	24.39	2.08	0.208
Total	104.48	10.94	1.094

Values 10^6: Calculated @ 10 per cent

NITROGEN FIXERS

Rhizobium and *Azotobacter* can be produced in conical flasks mounted on to a shaking platform offering an excellent way of producing bacterial biofertilizers on a small scale. The media used for their growth contain organic substrates like glucose, mannitol and yeast extract and inorganic salts like calcium carbonate, magnesium sulphate and potassium orthophosphate. Axenic mother cultures grown on agar slants are used to inoculate appropriate medium and put on the shaking platform. After the growth, individual flasks are pooled aseptically and mixed with carriers such as sterilized peat, lignite or fortified charcoal powder and dispensed into polythene bags and sealed. For large-scale production, fermentors with complete automation for control of temperature, nutrient level and pH are used. The mixing of the culture and carrier material is done in bulk and dispensed in presterilized containers.

Incorporation of efficient plant-specific species of *Rhizobium* right from sowing / seedling stage in nursery and subsequently in fields is highly desirable. Optimal results are obtained by the bacterization of seeds with the bacterial cultures, using 15% jaggery solution prior to their sowing in the root trainer cups or adding cell suspension prior to germination at nursery level. To ensure their viability in view of limited shelf life, it is desirable to produce desired quantities of fresh inoculum at site. However, this is feasible and cost-effective only where large-scale nursery operations are done for commercial plantation. To make such inoculations effective, they are inoculated in numerical superiority (1×10^8 cells/ml/plant) over the indigenous flora. Once their compatibility with local micro-flora is established, they continue to show their effect, unless the soil is exposed to wild fire, inundation due to flood or excessive use of pesticides. It is estimated that N_2-fixers fixes 20-30 kg of N_2 per hectare per annum, which is equivalent to the addition of 40-60 kg of urea.

PHOSPHATE SOLUBILISERS

Phosphorous (P) plays a crucial role in increasing crop yields. Unfortunately the plants use only 30 % of the applied P and the remaining gets converted to sparingly soluble/insoluble compounds such as tricalcium phosphate or phosphates of iron, aluminum and magnesium. Both bacteria and fungi have the ability to solublize these compounds by secretion of aliphatic and aromatic acids and enzymes such as phytases and phospholipases. These microbes are referred to as phosphate solubilising microorganisms (PSM) (Table 1.1). Different mycorrhizae also solublize insoluble phosphorous and finally helps in its mobilization. Application of mycorrhizae as biofertilizer has gained momentum in forestry sector, as they are known to confer several attributes to the plants including enhancement in growth of containerized seedlings and their subsequent better establishment and survival in field.

The use of PSM for effective utilization of P is yet to catch in a big way. Inoculations of PSM have shown positive response in crops such as rice. wheat, barley, tobacco sugarcane, potato and other vegetable and forage crops. PSM increases the efficiency of ground rock phosphate. Basic slag, bone meal and super phosphate applied to neutral and alkaline soils. Their inoculation at nursery stage affords stabilization with the root system so as to be compatible with the local micro-flora upon transplantation in the field. Integrated use of suitable microbial cultures along with low-grade rock phosphates can add 30-35kg P_2O_5 per hectare. India is perennially dependent upon large-scale import of phosphatic fertilizers and rock phosphate. Application of such microorganisms will have tremendous financial implications.

SWOT Analysis and Management

There are many industries in India, which are producing biofertilizers. The major strength of using such microbial inoculants is reduction in the use of chemical fertilizers at least by 20-30% and at the same time making the soil vital in an ecofriendly manner. Future approach in this area has to be carved out considering possible constraints, which may crop up in the large-scale application of this technology. The constraints can be:

(a) Non-availability of region specific microbes backed by an authenticated germplasm collection centre.

(b) Improper handling at production, distribution and consumer levels.

(c) Absence of strict quality control legislation and casual concern about quality management at the production level.

(d) Neglect of an appropriate research and development support.

FACTORS, WHICH LEAD TO LOW EFFECTIVE DEMAND

- *Lack of awareness among the consumers and promoters about the application of the merits of technology:* due to ready and quick response of chemical fertilizers, farmers are hesitant to use such microbial inoculants.

- *Off-the-shelf invisibility of biofertilizers:* knowledge of cropping pattern in an area or state at different time is required to make them timely available to the farmers which will also fetch good price, since demand and supply will be met through.
- *Lack of proper quality control measures:* The gullible consumer being duped by unscrupulous manufactures interested in making a fast buck. Production of bioinoculants starting from their mother cultures to starter cultures and mass multiplication followed by down stream processing requires sterile environment and is to be operated upon by skilled personal and assured titer value before packaging. Good quality of carrier material is also required for longer shelf life. Proper legislation with quality control and specifications laid down by Bureau of Indian Standards for certain products have to be strictly followed.
- Absence of Federal support in the form of a dependable germplasm collection of region specific microbes of agricultural importance. So that they function both as depositories and service centers for selected microbes. Such facilities are absolutely necessary at state level where microbes are made available from time to time to the manufacturers at nominal price.
- Information gap on microbe-microbe interaction and fate of bioinoculants in soil.
- Varying response of inoculants at different locations due to agro-ecological factors and the quality of the inoculum. Application of consortium of microbes can combat such situations.
- *Chemical fertilizers available at subsidized rates:* In spite of heavy subsidy on chemical fertilizers, which has increased substantially during the last decade, the cost of fertilizer nitrogen is prohibitive and the gap between demand and supply is widening. By 2000 AD the total fertilizer consumption is targeted at 240 lakh tons. of which nitrogen alone would be 140 lakh tons. To bridge the gap between demand and supply of chemical nitrogen, cheaper sources of nitrogen for sustaining agricultural production are needed.

A unit, capable of manufacturing about 1000 tons (5 million packets) of bacterial biofertilizers, supported by the required research and development wing, would require an investment of about Rs. 10 million. Land requirement for bacterial biofertilizer unit will be around

3000-5000 m^2 with 10-12 skilled and unskilled workers. Algal biofertilizer unit will require another investment of about Rs. 1 million and additional 7000 m^2 land with 10-12 skilled and unskilled workers. One can begin with the low investment algal biofertilizer unit and then take up bacterial biofertilizers and possibly microbial pesticides.

Marketing can initially be entrusted to agencies like National Biofertilizer Development Centre, National Dairy Development Board, Hindustan Lever Limited and State Department of Agriculture, who have the awareness as well as marketing channels developed for the agricultural inputs. This will appreciably cut down the initial investment and provide time to feel the market pulse. The selling price of a packet of biofertilizer sufficient to inoculate an acre is just double of the manufacturing cost the break even point is 40% with internal rate of return being 18% and 5 year break even period.

At present there are around 50 biofertilizer manufacturing units all over country of which roughly 10 units are in the private sector and rest are run by State Agricultural Department, chemical fertilizer industries and agro-industries in the Govt. sector. To fully utilize the production potential of these units, intensive efforts are to be made and substantial demand from farmers is to be generated.

Recommendations

- Entrepreneur will come forward to establish biofertilizer industry, only if, there is demand with good market potential and quick returns. Demand has been envisaged but technicalities have never been worked out properly. There exists a large consumer base, which is aware of the biofertilizer concept, and aggressive popularization campaign will accelerate the market penetration rate. This will be further facilitated if various Government agencies, industrial units and public and private R & D Institutions join hands to reveal economic and ecological benefits of the concept. A focused approach towards public/user awareness, quality control mechanisms for products, technical and fiscal support to small-scale businesses along with R& D support is required to develop entrepreneurship.
- There is a growing tendency to shift from chemical based agriculture to organic agriculture. Consumption of biofertilizers in North Eastern states has risen by 50% during the last 15 years. Depending upon the total area under different crops, a dedicated

popularization campaign is expected to generate an annual demand of 21 million packets of *Rhizobium,* 39 million packets of *Azotobacter* and 24 million packets of algal inoculants. Successful implementation and acceptance of biofertilizers at large scale will depend on timely availability of high quality products with sufficient shelf life, application at proper and precise time, use of specified equipments, with appropriate education and orientation to trainers, distributors and users.

— • —

Chapter 2

Potential Demand and Current Status of Biofertilizers Production in India

India has attained self-sufficiency in food grain production and has reached a production level of about 209 million tonnes during the year 1999-2000 with a buffer stock of 45 million tonnes currently. A consumption level of about 18.5 million tons of plant nutrients (NPK) and an average of 95 kg of nutrient per ha^{-1} has been attained during 1999-2000. The production of food grain is required to be increased continuously keeping in view of the increased growth in population and in the context the requirement of inorganic fertilizer will also be in the ascending order. Keeping in mind, the work on bio-fertilizers, the suitable alternate supplement for inorganic fertilizers has been activated in the country.

Although the production of *Rhizobium* was started in 1934, its first commercial production was initiated in 1956 at Indian Agricultural Research Institute, New Delhi and Agricultural College and Research Institute (now Tamil Nadu Agriculture University), Coimbatore. This was followed by the production of different types of biofertilizers in smaller quantities in State Agricultural Universities. Some state Governments also started production units to meet their local demand. The research activities on BNF have also strengthened in Agricultural Universities. In 1978, The Indian Council of Agricultural Research (ICAR) has started an All India Coordinated Research Project on Development and Use of Biofertilizers in 1983. Later in 1988-89, the Department of Biotechnology, Ministry of Science and Technology, Government of India initiated project on Technology Development and Demonstration of Biofertilizer for rice crop. As of 2001, more then 95 firms belonging to public and private sectors are involved in production of different types of biofertilizers at different parts of the country (Table 2.1).

Table 2.1. State wise number of biofertilizers plants in India

Zone	State	No. of Biofertilizers Plants
EAST		**12**
	Assam	2
	Bihar	3
	Orissa	1
	West Bengal	5
	Manipur	1
NORTH		**17**
	Haryana	2
	Himachal Pradesh	1
	Punjab	1
	Uttar Pradesh	12
	Delhi	1
SOUTH		**29**
	Andhra Pradesh	3
	Karnataka	11
	Kerala	1
	Tamil Nadu	13
	Pondicherry	1
WEST		**37**
	Gujarat	3
	Madhya Pradesh	7
	Maharashtra	24
	Rajasthan	3
ALL INDIA		**95**

(FAI, 2001)

The well-established and resourceful biofertilizers like *Rhizobium,* dominated the production earlier, but currently other microbial inoculants like *Azotobacter, Azospirillum, Acetobacter* and Phosphate solubilizing microorganisms (PSM) are being commercially produced and marketed. Recently, the production of Vesicular Arbuscular Mycorrhizae (VAM) has also been initiated in some production centers. Recently , Cadila Pharmaceuticals Ltd, Ahmedabad has launched two novel biofertlizers viz: Kalisena & Josh based on *Aspergillus niger* and *Glomus* VAM, respectively. Fertilizer Association of India has reported that the production of Biofertilizers in India increased from 2,005 MT in 1992-93 to 10, 000 MT in 1999-2000 (FAI, 2001). The current production of biofertilizer in the country its likely to reach 13, 000 metric tons (Table 2.2).

Table 2.2. Biofertilizer Production Scenarion in India

Year	Capacity ('000 T)	Production ('000 T)	Capacity Utilization (%)
1992-93	5.40	2.00	37.1
1993-94	6.12	3.08	50.3
1994-95	8.11	5.18	63.8
1995-96	10.68	6.69	62.6
1996-97	12.65	741	58.6
1997-98	16.36	7.10	43.4
1998-99	16.45	8.01	48.7
1999-2000	N.A.	10.04	-
2000-2001	N.A.	13.03	-

(FAI, 2001)

There has been a gradual increase in biofertilizer production over years, as the production in 1999-2000 was 5 times more than that in 1992-93. Since biofertilizer production is demand driven the capacity utilization of different biofertilizer production units depends largely on creation of demand for biofertilizers. However, an increase in the distribution of biofertilizers from 1992-93 to 1998-99 was noticed. The distribution of nitrogenous biofertilizer and Phosphobacteria biofertilizer is almost equal. Depending on the source of funding, biofertilizer production units located in different states are categorized as (i) those financed through Government of India (GOI) and (ii) those financed by other sources. So far, 67 biofertilizer production units have been established with the GOI support. These units share above 9 thousand tons of the total installed capacity and 6 thousand tons of the total biofertilizer production. Besides 28-biofertilizer production units, other than GOI financed units, with a total installed capacity of 8 thousand tons produce nearly 4 thousand tons of biofertilizer annually. Private entrepreneurs, NGO's and some of the Agricultural Universities establish these units. The installed biofertilizer (*Rhizobium, Azotobacter, Azospirillum* and P Solubilisers) production capacity of about 16,450 tonnes is much lower than the potential demand, which is estimated to be 3.4 lakh metric tons as observed by Biotech Consortium India Ltd., Delhi. This indicated substantial growth of this microbial industry and eventually entrepreneurship.

Table 2.3. Present status of biofertilizer technology in Developing Countries.

Rhizobium	Azospirillum	Azotobacter	Azolla & BGA
BANGLADESH			
Important Efficient Strains isolated for soybean and other important legumes Wide Testing in farmers field, commercial production and distribution needed	Still under research	Under study	Not presently important Limited research A temperature Resistant Strain of Azolla is yet to be found
INDIA			
Important Need of good strain and quality testing More field demonstrations needed.	Response +ve with sone cereal crops more field demonstrations needed	Important, found effective for some cereals & vegetables	important in some coastal and NE states for Paddy
INDONESIA			
Important	Study in progress	Not much work done	Not important
NEPAL			
Important Testing in farmers field & demonstrations are needed	Very little work done	Limited research conducted	Not presently important

Contd...

Table 2.3. Contd...

PAKISTAN Important Wider testing for commercial production essential	More work needed with wide field demonstration	Study in progress & result encouraging	Not presently important
SRILANKA Important strains of soybean, cowpea and groundnut Isolated Wider field testing essential	Study has been initiated	Not much work carried out	Needs further testing
THAILAND Important Wider testing and demonstration needed Commercial Production facility exists. Attention to quality required	Still under research	Limited research carried out	Azolla is a potential source but lack of water during dry summer is a c

Field trials and cooperative agreements with the private sector exist. As a result, the quantities of inoculants produced by Department of Agriculture have increased from 5 to 211 tons and their use has increased from 3.36 to 203.28 tons during 1977-1995.

Table 2.4. Rhizobial inoculant production and utilization in thailand

Year	Production	Utilization
1977	5.00	3.36
1978	10.59	11.55
1979	7.42	5.77
1980	4.92	5.64
1981	7.48	7.36
1982	6.58	6.64
1983	14.36	13.15
1984	36.16	33.79
1985	48.77	46.51
1986	78.00	74.78
1987	81.63	79.79

Contd...

Table 2.4. Contd...

1988	140.07	136.23
1989	134.27	125.03
1990	126.35	117.67
1991	73.78	72.03
1992	94.44	92.81
1993	179.78	171.53
1994	193.39	176.15
1995	211.41	203.28

Myanmar

The economy of Myanmar is largely based on agriculture, with about 76% of the population depending on agriculture for their lively hood. Peanut and food legumes occupy about 2.7 million ha of the 8.9 million ha of cultivated land. Black gram, Green gram, Pigeon pea, chickpea, soybean and cowpea are the most important legumes. Food legumes play a major role in such cropping systems as rice-legume; sesame-legume, and in mixed cropping. Food legumes are produced mainly for domestic consumption, with the surplus being exported. Nitrogen supply to these crop is supplemented by the application of urea, which is the only chemical fertilizer available as a N source in Myanmar. However, the urea available is insufficient even for the rice crop. Biological nitrogen fixation (BNF) was identified as a possible alternative N source. The *Rhizobium* production was recorded maximum during 1982 and 1986 (114 and 103 tonnes respectively) but declined to 27 tonnes in 1995.

Table 2.5. Rhizobium inoculant production in Myanmar

NO. OF PACKETS							
Year Nut	Ground Bean	Butter	Chickpea Gram	Blank bean	Mung Legumes	Other	Total
1978	0	0	50	0	0	0	50
1980	33,272	2,450	1,26,345	0	0	0	1,62,067
1985	4,07,850	53,040	1,55,370	0	0	1,760	6,18,020
1990	1,23,973	12,162	73,680	22,948	19,439	0	2,52,202
1995	27,725	0	65,259	39,777	44,812	3,335	1,80,908

African and Latin American Countries

In the East and Southern Africa region, legume inoculants are produced from commercial purposes in Kenya, Malawi, Uganda, Tanzania, South Africa, Zambia and Zimbabwe (Table 2.6). A total of 154 tons of legume inoculants are currently produced in the region for a wide range of leguminous crops, with East African countries producing 5,800 kg. Currently Rhizobium inoculant production takes place in Kenya in the Network of Microbial Resources Centre (MIRCEN) laboratory at the University of Nairobi and MIRCEN collaborating laboratories at Makerere University and the Madhivan Sugar Company in Jinja, Uganda. The produces of different brands of legume inoculants include research institutes, universities and private entrepreneurs. Current needs of inoculants are met through the importation of 11 tonnes /yr from the two producers in Uganda. Inoculants in Zambia and Zimbabwe are produced by Government facilities essentially to stabilize and service emerging soybean industries. The production of these Rhizobial inoculants averts the dependence on inorganic N fertilizer.

Table 2.6. Summary of legume inoculant production in different african countries

Country	Quantity (kg)	Carrier inoculants
Ethiopia	Peat	Clovers, Desmodium, Sesbania, Lucerne, Stylosanthes
Kenya	Filter mud	Lablab, Soybean, Beans, Alfalfa, Desmodium, Pigeon pea, Lucerne
South Africa	Irradiated peat	soybean, Groundnut, Cowpeas, Lucerne, Clovers, Beans, Peas, Lotus, Desmodium, Medicagos
Uganda	Peat	Soybean, Beans, Lucerne, Groundnut
Zambia	Peat	Soybean 90%, Lucerne, Bean
Zimbabwe	Sterile bagasse	Soybean 90%, Beans, Peat, Lentil, clove, Lucerne, Desmodium, Groundnut, Stylosanthes, Crotalaria.

Among the Latin American countries, Brazil is the second largest producer of Soybean in the world with 30 million tons in 1999. This was made possible with research and dissemination of the rhizobial inoculation technology in the use of legume production in the country.

Of the 150 efficient strains selected for grain and forage legumes, 20 strains were for soybeans. The private sector in Brazil produces approximately 10 million packages (200 g per package) of inoculant per annum, of which 95% is for the soybean crop. Since 1995, Argentina, Brazil, Paraguay and Uruguay, at a Governmental level, have defined research priorities of regional interest and basic criteria and principals concerning sterility of carriers, concentration of viable Rhizobial cells in manufactured products, registration and quality control of these products and establishment of centers of reference and standards, etc. within the framework of the common market of inoculants bartered through the Latin American Common Market.

—•—

Chapter 3

Biofertilizers Technology

The term "Biofertilizer" is popular misnomer. It refers to living organisms, which augment plant nutrient supplies in one way or the other. In the strictest sense, real biofertilizers are the green manures and organics (materials of biological origin, which are added to deliver the nutrients contained in them). But, we will restrict biofertilizer to:

1. Carrier-based inoculants containing cells of efficient strains of specific microorganisms (mainly bacteria) used by farmers for enhancing the productivity of the soil either by fixing atmospheric Nitrogen (N) or by solubilizing soil Phosphorus (P) or by stimulating plant growth through synthesis of growth promoting substances.
2. Blue green algae or Cyanobacteria .
3. Mycorrhizae

Biofertilizer may be broadly classified into Nitrogen Biofertilizers (NB) and Phosphate Biofertilizers (PB). Nitrogen biofertilizers are now under use. Phosphate biofertilizers are also in use but on a lesser scale and are yet to be standardized.

IMPORTANCE AND CONTRIBUTION OF BIOFERTILIZERS IN AGRICULTURE

Eighty per cent of the population in India depends on agriculture. Majority (75 %) of the farm holdings are small and marginal farmers having less than one hectare and they till 54.6 % of the total farm holdings, while small farmers having 1-2 hectares holding constitute 18 %. It is very difficult for them to purchase and use recommended fertilizer doses at current prices. They need to exploit other less expensive nutrient sources to the maximum. In order to raise their incomes and living standards, these landholders must maximize crop productivity per unit area in the most cost effective manner. Biofertilizers, based on renewable energy source are a cost effective supplement to chemical fertilizers and can help to economic on the high investment needed fertilizer use as far as N and P are concerned.

Biofertilizers are known to make a number of positive contributions in agriculture for example:

1. They supplement fertilizer supplies for meeting the nutrient needs of crops.
2. They can add 20-200 Kg N/ha (by fixation) under optimum conditions and solubilize/mobilize, 30-50 Kg P_2O_5/ha.
3. They liberate growth promoting substances and vitamins and help to maintain soil fertility.
4. They suppress the incidence of pathogens and control diseases.
5. They increase crop yield by 10-15%, N-fixers reduce depletion of soil nutrients and provide sustainability to the farming system.
6. They are cheaper, pollution free and based on renewable energy sources.
7. They improve soil physical properties, tilth, and soil health in general.

Associative and free living microorganisms are believed to contribute to sustainability of flooded rice production system. Approximately 50% of the N requirement of flooded rice is met from the pool of soil N that is believed to be maintained through BNF by associative and free living microorganisms. Contributions from symbolic N_2 fixation of the order of 160 Kg N/ha have been reported for sugarcane.

—•—

Chapter 4

Rhizobium

The root nodulating bacteria invented by Hellriegal and Wilfarth (1988) was first isolated and named as *Bacillus radicola* by Beijerinck in 1988. The generic name was changed later to *Rhizobium*. *Rhizobium* is one of thc important nitrogen-fixing bacteria, which helps the legume to maintain the soil fertility by way of their symbiotic association and nitrogen fixation. A legume plant having effective root nodules not only can meet its own nitrogen requirement but can also enrich the soil nitrogen content and thus improving soil fertility and sustainability. Legumes are cultivated widely in different parts of India. The practice of seed inoculation with *Rhizobium* is popular among farmers. The N_2 fixation by different legumes is estimated to be 100-300 kg ha^{-1} in pea and 53-85 kg ha^{-1} in cowpea.

CLASSIFICATION

1. Family : *Rhizobiaceae.*
2. Genus : *Azorhizobium* for stem nodulation *(Sesbania rostrata)*
   : *Bradyrhizobium* (Soyabean, lupin, cowpea miscellany i.e. cowpea, greengram, redgram, chickpea, groundnut).
   : *Rhizobium* (Pea, lentil, bean, lathyrus, berseem, lotus).

MORPHOLOGY

Unicellular, cell size less than 2μ wide. Short to medium rod, plemorphic, motile with peritricus flagella, Gram negative, accumulate poly β-hydroxy butyrate granules.

PHYSIOLOGY

Nature	:	Chemoheterotrophic, Symbiotic with legume,
C-source	:	Supplied by legume through photosynthates, monosaccharides, disaccharide.
N-source	:	Fixed from atmosphere
Respiration	:	Aerobic
Growth	:	Fast (*Rhizobium*), Slow (*Bradyrhizobium*)
Doubling time	:	Fast growers (2-4 hours) Slow growers (6-12 hours)
Growth media	:	YEM/YEMA (Fred).

CONTRIBUTION

1. Direct contribution of N symbiotically with legume:

Alfalfa	:	100-200 kg N/ha
Pigeon pea	:	168-200 kg N/ha
Clover	:	100-150 kg N/ha
Chickpea	:	85-110 kg N/ha
Cowpea	:	80-85 kg N/ha
Groundnut	:	50-60 kg N/ha
Lentil	:	90-100 kg N/ha
Greengram/ Blackgram	:	50-55 kg N/ha
Pea	:	52-57 kg N/ha
Soybean	:	60-80 kg N/ha

2. Residual nitrogen benefit for the succeeding crop.

RECOMMENDED FOR LEGUMES

Pulses : Chickpea, pea, lentil, black gram, green gram, cowpea, pigeon pea.

Oil seeds : Soybean, groundnut,

Fodders : Berseem (Egyptian clover), Lucerne.

Increase in Yield: 10-35 %

Common Name of Crops	Rhizobium sp. (with cross inoculation group)
FAST GROWING RHIZOBIA	
Lucerne, Fenugreek	*Rhizobium meliloti*
Egyptain, clover, Clover	*R. trifolii*
Khesri (Lathyrus), Lentil, Pea, Common vetch	*R. leguminosarum*
Bean, Kidney bean, French bean	*R. phaseoli*
Chickpea, Subabul	*Rhizobium* sp.
SLOW GROWING RHIZOBIA	
White lupines, Lupines	*R. lupines*
Soybean	*R. japonicum*
Sun hemp, Cluster bean, Peanut, Lespedeza (Kudzu), Jack bean or Sward bean, Lablab, Horse gram, Moth bean, Black gram, Green gram, Cowpea, Pigeon pea	*Rhizobium* sp.

Note: Several *Rhizobium* strains of cowpea miscellany may be fast growing.

To become familiar with legumes in the field, examine their nodules, isolate rhizobia from nodules, authenticate isolates and preserve the isolates by desiccation on ceramic beads.

Procedure

1) Identify legumes in the field, collect nodulated specimens and preserve nodules.
2) Examine nodules and bacteroids under the microscope.

3) Surface sterilize nodules and isolate rhizobia on differential media.
4) Perform Gram stain and reisolate on differential media.
5) Store isolates on slants.
6) Surface sterilize and pregerminate seeds for authentication.
7) Plant and inoculate seedlings for authentication.
8) Examine plants periodically for nodulation.
9) Terminate experiment, examine nodules and reisolate.
10) Prepare broth culture of authenticated isolate for desiccation on beads.
11) Prepare bead storage vials
12) Impregnate sterilized beads with broth culture of rhizobia
13) Regrow rhizobia stored on beads

(a) Recognizing legumes and identifying them in the field

Note the main similarities among all legumes in their compound leaves and the seed placentation in pods. However, in many Acacia species (e.g. *Acacia auriculaeformis, Acacia mangium, Acacia Koa)* the compound leaves are only formed and seen in seedlings. The compound leaves are replaced by phyllodes as the plants mature. Compound leaves are also possessed by numerous non-legumimous families such as: Bignonoacae (e.g. Jacaranda, Spathodea); Caprifoliaceae (e.g. Sambucas mexicana var. bipinnata); Solanaceae (e.g. Lycopersicon, Solanum tuberosum), Passifloraeae (e.g. Passiflora spp.) Familiarize yourself with basic characteristics of each subfamily as outlined earlier. Learn to identify legumes in the field and become familiar with the appearance of the most common agricultural legumes in your area.

(b) Recovering nodules in the field

Identify plants of several legume species in the field and select one representative of each for sampling. With a spade, describe a circle with a radius of approximately 15 cm around the plant and cut out this section to a depth of at least 20 cm or deeper. Still using the spade, slowly lift out the clump. Carefully remove the soil from the root material with your hands. Avoid detaching secondary roots from the plant as nodules may be found on the lateral roots as well as the taproot.

Carefully place the whole plant into a plastic bag. If the legume has seed, collect the seeds and store them in the refrigerator for the authentication test.

In the laboratory, place a sieve of an appropriate size and mesh under each root sample. Carefully wash the roots under a gentle stream of water from a tap or a hose.

The distribution of the nodules on the root system is dependent on the species of host plant and *Rhizobium* strain as well as soil structure and composition of the soil.

(c) Preserving nodules

Fresh nodules may be stored in the refrigerator overnight. Do not freeze nodules as the forming of ice crystals may rupture and kill the bacteroids. Frozen nodules may, however, be used for serological typing. For long-term storage, desiccation in glass vials is recommended.

(d) Examining nodules and bacteroids

Note the shape and size of the nodules recovered from the collected plants. Nodule size and shape vary with *Rhizobium* and host plant species. Large round nodules may be found on cowpea and soybean plants. *Leucaena* and *Acacia* are among legumes, which do not have round nodules.

Carryout thin sections of nodules with a razor blade and float them on a drop of water on a microscope slide; use a cover glass and examine under low power (10 x) and high power (40 x) objectives.

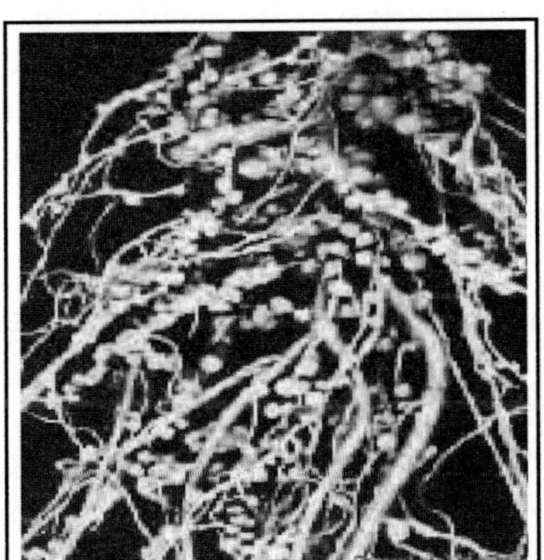

Nodules

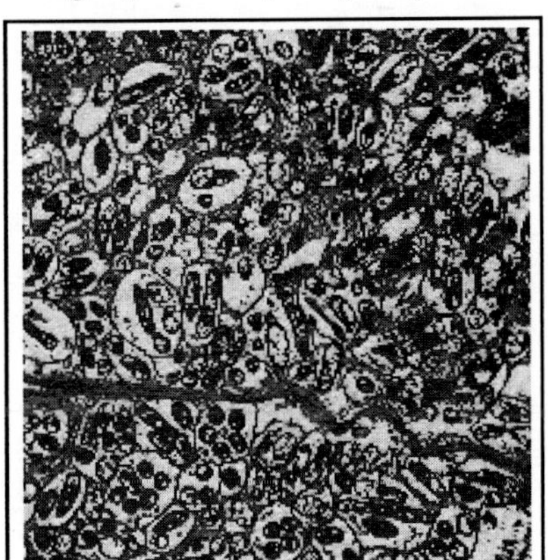

Bacteroids

An Active N-fixing nodule contains a protein called leghaemoglobin. Its presence in the nodule can be noted by the characteristic pink, red, or brown coloration. Active nodules may also be black. These are not very common. Black nodules have been reported on *Lablab purppureus, Dolichos biflorus*, and *Vigna unguiculata* when inoculated with some strains of rhizobia.

Senescent nodules are usually grayish green. Nodules with a green exterior may also be found when nodules on the soil surface are exposed to sunlight. This exterior green color is due to chlorophyll development on the cortical region of the nodule. Most ineffective rhizobia cause nodules with white interiors which lack leghaemoglobin.

Gently rub the cut surface of a nodule on a clean microscope slide to make a smear. Allow the smear to air dry and then pass the slide through a flame. Cool the slide and stain the smear with dilute carbol fuchsin for 10-20 seconds. Wash in water, blot off excess moisture, and air dry. Examine under the oil immersion objective. Note the difference in morphology between the "bacteroids" in this smear and bacteria of the same rhizobial species grown in pure culture. Note the size and shape of the bacteroids compared to the rod forms found in pure culture.

(e) Isolating Rhizobium from a nodule

Wash roots thoroughly to remove soil. About 10 nodules are collected from each plant. Separate the nodule from the root by cutting the root about 0.5 cm on each side of the nodule. The root appendages can be used to manipulate the nodule with forceps and reduce the risk of damaging the nodule.

Immerse undamaged nodules for 5-10 seconds in 95% ethanol or isopropanol (to break the surface tension and to remove air bubbles from the tissue); transfer to a 3% solution of calcium hypochlorite, and soak for 4-6 minutes. Rinse in five changes of sterile water using sterile forceps for transferring. Forceps may be sterilized quickly by dipping in alcohol and flaming. Utilize sterile glass or plastic Petri dishes as containers for the ethanol, calcium hypochlorite and water.

Alternatively, nodules may be placed into an Erlenmeyer flask (125 ml). The sterilizing and rinsing fluids may be changed as required, leaving the nodule in the flask each time.

An acidified mercuric chloride solution (0.1 % w/v) or a solution of hydrogen peroxide (3 % v/v) may be used for sterilizing nodules.

However, mercuric chloride is highly toxic and hydrogen peroxide is expensive, making calcium hypochlorite is the preferred choice. When hydrogen peroxide is used, 5 - 6 rinses with sterile water may be omitted.

Desiccated nodules must be rehydrated before sterilizing. Place nodules into a small beaker with clean cool water and leave in the refrigerator to imbibe overnight. A one-hour soaking at room temperature is satisfactory for nodules, which have been desiccated for only a short time.

Crush the sterilized nodule with a pair of blunt tipped forceps in a large drop of sterile water in a Petri dish. Alternatively, the nodule may be crushed in a sterile test tube with sterile glass rod. Streak one loopful of the nodule suspension on yeast-mannitol agar (YMA) plates containing Congo red or bromthymol blue

The primary isolate may be streaked in one continuous motion.

Deposit culture on agar with the inoculation loop then streak out to 1. Resterilize loop, and cool by touching the agar surface near the side of the Petri dish, cool, streak from 1 to 2. Repeat the procedure until 4 is reached.

The *needle method* of isolation is especially useful with freshly harvested nodules 2 mm or larger in diameter. The surface sterilized nodule is first placed on a small piece of sterile filter paper (2 cm x 2 cm) in a sterile Petri dish. A new piece of filter paper should be used for each nodule. The same Petri dish can be used for several nodules. Wash the nodule well with water. Dip the blunt tipped forceps into 95% alcohol and flame momentarily. While holding the nodule in the forceps and resting the nodule on sterile filter paper, quickly slice off a small section with a flamed, hot scalpel. Still holding the nodule in the forceps on the filter paper, insert a sterile inoculation needle, with a 1 mm into the cut surface to load the loop with inoculum. Streak directly onto YMA plates containing Congo Red and onto a YMA plate containing bromo thymol blue (BTB).

The needle method requires steady hands. Resting the elbows of both arms on the tabletop while manipulating the forceps and needle is helpful.

For small nodules, the tip of the inoculation needle may be bent over and doubled up for a length of 1 mm or even left unbent. The needle itself should not be longer than 5 cm.

Another method consists of *serially diluting* the nodule bacteroids suspension and then pour plating it. With a sterile Pasteur pipette, place two drops of water separately into four empty sterile plastic Petri dishes marked A, B, C, and D. Crush the sterilized nodule in a sterile Petri dish or test tube.

Flame the transfer loop and cool it drop-1 of dish-A, then transfer the bacteroids suspension from the crushed nodule to drop-2 of dish-A and mix.

Next, flame the loop, cool it in drop-1 of dish-B, and transfer one loopful from drop-2 of dish-A to drop-2 of dish-B and mix. Continue until drop-2 of each dish has been inoculated and mixed with the diluted nodule suspension of the previous one.

Pour 15-20 ml liquid YMA (48°C) to the inoculum in each dish. Ensure mixing by gently moving the covered dish first clockwise and then counter-clockwise on the tabletop. Allow three full circles for each movement. Continue mixing by moving the dish from the left to the right and from the right to the left three times. Then, without pausing, move the Petri dish forward and backward and backward and forward, also three times. Allow the agar to set, before incubating. Invert the plates during incubation.

PROCESS

- Wash roots thoroughly
- Select pink and healthy nodules.
- Collect nodule from plants without damaging it.
- Immerse undamaged nodule in 95% ethanol for 5-10 seconds
- Sterilize nodule surface by acidified mercuric chloride solution (0.1% W/V).
- Rinse nodules 5-6 times in sterile water
- Crush the sterilized nodule with forceps under large drop of sterile water in a petridish.

AUTHENTICATION AND STRAIN SELECTION

After isolation, the strains are made pure and authenticated. For this, following tests are done.

1. Growth in YEMA (Yeast Extract Mannitol Agar) for observing its colony character.

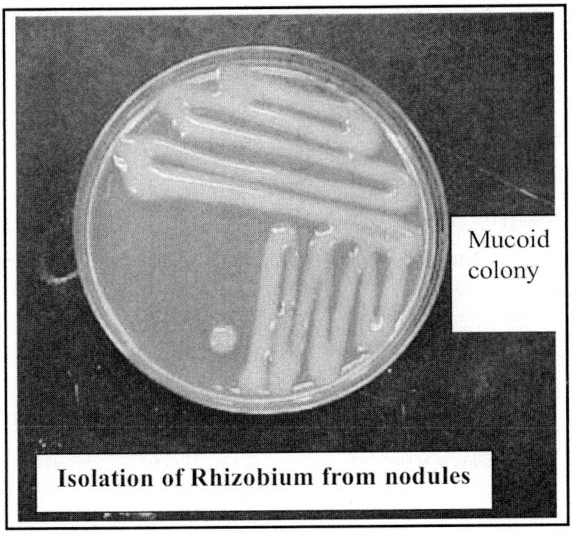

2. Growth in CRYEMA (Congo Red Yeast Extract Mannitol Agar) for confirming its growth.
3. Growth in BTB (Bromothymol Blue), YEMA medium for observing whether it is fast-grower or slow-grower.
4. Growth in Hoffer's alkaline medium (for confirming as to whether it is Rhizobium or Agro bacterium).
5. Gram reaction.
6. Nodulation test (for confirming its effectiveness).

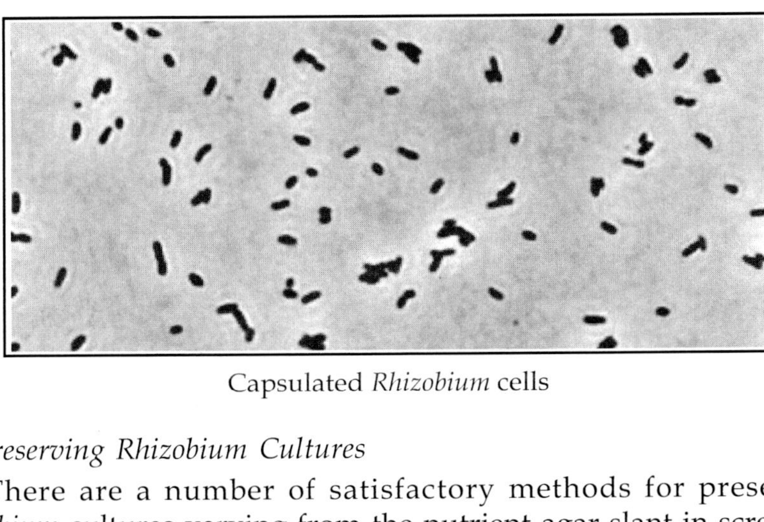

Capsulated *Rhizobium* cells

(h) Preserving Rhizobium Cultures

There are a number of satisfactory methods for preserving *Rhizobium* cultures varying from the nutrient agar slant in screw-cap tubes to the desiccated form on porcelain beads, from the lyophilized (freeze-dried) to frozen liquid suspension under liquid nitrogen. The choice of method will depend on facilities and experience. The porcelain bead method is recommended for laboratories with limited resources.

In preparing for storage on beads, inoculate a loopful of culture from a YMA slant into 3 ml of sterile YM-broth and incubate to maximum turbidity on a rotary shaker.

Place 20-30 ceramic beads (washed and oven dried) in a screw-cap test tube, cover the mouth of the tube with foil, and sterilize in the oven for 1-2 hours at 160-170 °C. Prepare storage tubes using 6-7 g silica gel and sufficient cotton or glass wool to keep the silica gel in place. The rubber-lined caps for the tubes must be autoclaved separately, then dried in an oven at 80-90° C.

The glass wool may be oven sterilized in the storage tube with the silica gel. When cotton is used, it should be autoclaved in small balls in a foil covered beaker. These cotton balls should be of a suitable size to facilitate easy aseptic transfer to the storage tube with forceps. Residual moisture is removed in the oven at 70-80° C before transferring it aseptically to the sterile storage tubes.

Transfer the sterilized beads aseptically to the broth culture in the tubes and replug. Soak for 1-2 hours then invert the tube and allow the excess broth culture to soak into the cotton plug. Transfer the beads impregnated with rhizobia into the storage tube aseptically, replace and tighten the screw caps securely.

In regenerating a culture, inoculate YM broth one or two beads. These are easily speared from the storage tube using a sterile needle with a slight hook. A week or more may be needed to obtain visual signs of growth. Once the broth becomes turbid, loopful should be streaked on presumptive test media to check for retention of purity. Subculture from the broth onto YMA slants as desired.

Preparation of Broth (Mother Culture)

Following the selection of suitable strain, it will be inoculated into yeast extract mannitol liquid medium. The incubation is to be made at 28°C ± 2 on a rotary shaker for 2-7 days depending on the type of rhizobia.

Fermentation

After the medium is formulated and sterilized in fermenter, it is inoculated with starter culture. The amount of inoculum culture to be transferred in to the fermenter vessel depends on the size of the fermenter. The broth is continuously aerated by forcing sterile air through porous stainless steel sparger Aeration requirement of rhizobia varies from species to species depending on fermenter capacity. When the number of rhizobia in broth will reach 10^8 to 10^9 cells/ml, it will be ready for mixing with the carrier.

A large amount of inoculum reduces the time required to reach maximum viable numbers and, therefore, reduces the chance and effects of contamination. Inoculum levels generally vary from 0.1 to 1%, sometimes it is 5%, but generally should provide $10^6 - 10^7$ rhizobia/ ml culture medium at the beginning of fermentation. "Fast-growing strains" having a mean generation time of approximately 4 hours and reach maximum viable number in 24-52 hours compared to 78-156 hrs for "slow-growing strains" which have mean generation time of about 6-12 hours. Maximum numbers may be maintained for a further 24-36 hour, depending on the growth conditions although the proportion of viable cells will decrease. This is illustrated with an example below:

30 ml of fast growing Rhizobium having population of 2×10^9 cells was inoculated as starter culture to 3000 ml liquid medium at 2:30 p.m. on a day for fermentation. The meantime of *Rhizobium* is 3 hours. When will the population reach 10^9 cell/ml.

When 30 ml of 2×10^9 cells/ml at 1% inoculum level was added to 3000 ml liquid medium, the cell number will be:

$$\frac{30 \times 2 \times 10^9}{3000} = 2 \times 10^7 \text{ cells/ml}$$

Here the Mean Generation Time (mgt) of the strain is 3 hours.

Therefore multiplication will be as follow:

Time	Cell Number	Time	Cell Number
0 hour	2×10^7	12 hours	32×10^7
3 hour	4×10^7	15 hours	64×10^7
6 hour	8×10^7	18 hours	128×10^7 or 0.28×10^9
9 hour	16×10^7	21 hours	256×10^7 or 2.56×10^9

Although the cells attained 10^9 populations after 18 hours, adequate cells (2.5 x 10^9) were received 21 hours after inoculation. If inoculation is done at say 2.00 pm (1400 hrs), the broth is ready at 11.00 a.m. the next day.

MASS PRODUCTION

Production Under Non-sterile System

The steps involved are illustrated in the non-sterile production system usually means the use of unsterilized carries which are normally held in the dried form after grinding to avoid the natural population of organisms to have an opportunity to multiply before the Rhizobium broth is added. The mixing procedures vary from simple hand mixing on the floor, to ribbon or paddle type mixers.

- Tested seed culture is transferred aseptically either to the flask (1 % v/v inoculum rate), kept in rotary shaker or to the fermenter of suitable volume.
- In case of fermenter, the broth is aerated with a jet of sterile air @ 10-12 liters /hour. The temperature needs to be maintained a 28°C±2.
- The broth is harvested for mixing with the sterilized carrier , when a cell count of about 10^9 cells/ml is achieved. Incubation will be done for 3 to 5 days.
- Mixing of broth with the carrier is done under aseptic conditions either manually or mechanically, using ribbon or paddle type batch mixers. The preparation of broth added to the carrier is determined by moisture holding capacity of carrier. Enough broth should be added to occupy 40 to 50% of total water holding capacity of carrier. The carrier should be finely powdered to pass through 100-200 mesh sieves. It should be neutral. In case of peat and lignite, pre-treatment is made with the addition of $CaCO_3$ for neutralization.
- After mixing, the impregnated carrier is allowed to cure for 2 days to one week at 28-30°C for further multiplication of Rhizobium in the carrier.
- Finally the impregnated carrier is packed in the polythene packets, stored in cool and dry place at a temperature of 15°C but not exceeding 30°C.

- At every stage of production, e.g. at fermentation and curing, quality control as per BIS manual should be strictly followed.

Production Under Sterile System

Various steps in production under sterile system are as follows :

- The primary culture is made in a conical flask with an emptying device at its bottom placed on an orbital shaker at room temperature for 3 days.
- When fermentor is emptied through a medium dispenser and a sterile rubber pipe fitted with disposal needle.
- The carrier (120 g) taken in autoclavable polypropylene bags are sterilized for two consecutive days for 60 minutes each and the bags are sealed immediately after autoclaving.
- Once the sterilization is over, the bags are kept in the storeroom before inoculation during 3 days to allow the gases to escape.
- The carrier bag is quickly wiped with alcohol and the broth is injected straight from the fermenter into the sealed bag. About 80 ml are injected per bag.
- The injection hole is sealed with adhesive tape.
- An automatic blender blends the inoculant packets.
- The inoculant packets are moved to the incubation room and kept there for at least one week.
- After labeling, the inoculum is ready to be dispatched for sale.
- At every stage of production, the quality testing as per BIS manual should be strictly followed.

Advantages and disadvantages of using sterile and no-sterile carrier are summarized in the box given below. It is interesting that while Liphatech Co., USA (the old Nitragin Company) produce biofertilizers in non-sterile system, Liphatech, France produce *Rhizobium* under sterile system. The same is also observed in Thailand where the Department of Agriculture is using the non-sterile system while a commercial organization is following the sterile system. Choice of the system depends on local facilities as well as capacities.

ADVANTAGES	DISADVANTAGES
Sterile Carrier System	
☆ Used in small medium scale operation.	☆ Requires sterile carrier.
☆ Dilution of broth means reduced investment in fermentation vessels, media and supplies to produce rhizobial broth.	☆ Need to sterilize indigenous materials or purchase pre-sterilized, pre-packaged material.
☆ High quality product with long shelf life.	☆ Extra expenses and problems associated with importing this prepared material.
☆ Production schedule not as rigid due to longer shelf life.	☆ Need method for aseptic dilution with sterile water.
☆ Ease of quality control measures.	☆ More intensive handling needed. Broth must be introduced and mixed into individual packets.
☆ Pre-packaged, pre-sterilized materials are convenient.	☆ Aseptic handling requires trained personal.
Non-Sterile Carrier System	
☆ Used in medium to large-scale production systems.	☆ Broth dilution technique cannot be used and larger fermentation capacity required.
☆ Incorporating rhizobial broth into carrier material is a simple process.	☆ Need more supplies for producing large quantities of broth.
☆ Materials handled in bulk.	☆ Shorter shelf life of product and thus rigid production schedule.
☆ Indigenous materials can be used.	☆ Refrigeration of product to extend shelf life is expensive.
☆ Processing of indigenous materials reduced since sterilization not necessary.	☆ Quality control more difficult. Enumeration or rhizobia requires facilities for conducting plant infection test.
☆ Reduced need for specially trained staff.	

Product Formulation Technology

Based on the physical nature and carrier materials used, various types of biofertilizers are manufactured by different producers. These are carrier-based or agar based broth culture or dried culture. Some new developments in biofertilizer product formulation are:

- Freeze-dried inoculants
- Rhizobium paste
- Granular inoculant
- Pelleting
- Polyacrylamide entrapped rhizobia
- Pre-coated seeds etc. These appear more promising for tropical legumes.

Agar Based and Broth Culture

Agar based inoculants have become outdated except for small-scale experiments. Broth cultures are in vogue in Europe and in USA. Liphatech, France is marketing liquid legume inoculant 'BIOLIDOZ' (with shelf life of 24 months). The limitation of liquid broth is that when broth is directly used for inoculation, large quantities of fresh broth are necessary.

Frozen Concentrates

It is a concentrated broth culture of Rhizobium and has excellent storage properties. In the Netherlands, these have been distributed successfully. Effort was made to commercialize them in Australia, but it did not succeed, as it could not survive better on the seed. In India, IARI and BAIF have tried in but product was not accepted by farmers.

Granular Inoculants

Granular inoculants are prepared in New Zealand by binding a finely pulverized peat inoculant to coarse sand using an adhesive. It has the advantage of uniform; distribution through drill applicators but more R & D work is needed on this line of products.

Granules

In France, three inoculants - Polyacrylamide (PER), an alginate (AER) and xanthan/ carobgum (XER) inoculants have been used for inoculating soybean. The system has yet to be standardized. This also holds for porous gypsum granules, which are prepared by extruding gypsum paste through small circular openings and sprayed on broth cultures of rhizobia mixed with milk powder and sucrose.

Carrier-Based Inoculants

Liquid inoculants are mixed with carriers like peat, lignite and charcoal etc. Although theses have a shorter shelf life, still they are found most suitable for commercial use. Good carriers can support one billion (10^9) cells/g when the inoculant is fresh and maintains 100 million (10^8) cells/g for six months after production.

Estimation of Biological Nitrogen Fixation (BNF)

Common techniques used for estimating BNF in the greenhouse/ field are:

- Acetylene (C_2H_2) reduction
- ^{15}N gas incorporation
- ^{15}N isotope dilution
- Natural abundance of ^{15}N
- N-difference method
- N-balance method

Acetylene Reduction Assay (ARA)

The ARA makes use of the fact that the enzyme nitrogenase, which is responsible for nitrogen fixation, can reduce acetylene (C_2H_2) to ethylene (C_2H_4) as well as N_2 to NH_3. Acetylene and ethylene can be easily and rapidly analyzed by a gas chromatograph and hence the amount of N-fixed can be estimated. Number of strains can be evaluated on the basis of ARA estimation. For example, Pigeon pea Rhizobium strain G 3 gave ARA of 6.19 μ mol acetylene reduced/h/ plant and a shoot weight ratio 2.3. Thus ARA assay is a good tool for selecting efficient strains.

^{15}N Gas Incorporation

Nitrogen gas that is labeled with the stable isotope ^{15}N place in the atmosphere of a closed ^{15}N chamber in which the test plants are growing. The amount of nitrogen fixed is calculated by the amount of ^{15}N incorporated into the plant. One of the first applications of mass spectrometry to the use of stable isotopes in biology was the analysis of the incorporation of ^{15}N gas into culture of *Azotobacter vinelandii*. However, the high cost of ^{15}N is discouraging in many cases.

^{15}N Dilution

The principle of this method is that N_2 fixing plant growing in a medium enriched by ^{15}N fertilizer contains less ^{15}N in the plant than does a non-fixing plant growing into the same medium. The scientists from Department of Agriculture, Thailand measured the nitrogen fixation capacity of soybean by such a method.

Natural Abundance of ^{15}N

This method of calculating percent N derived from atmospheric N_2 is based on the same concept as that for the use of ^{15}N dilution and makes use of the differences between the abundance of ^{15}N in the soil and that of the atmosphere. Use of this method requires access to an isotope ratio mass spectrometer.

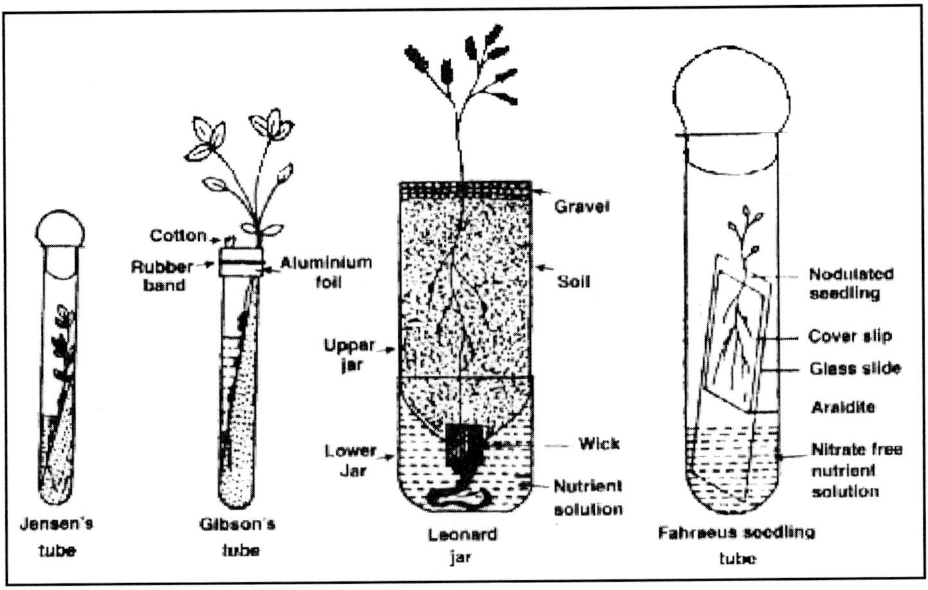

Testing of Rhizobium for N fixation ability

N- Difference Method

This classical method has generally been used by agronomists in fields, where the N_2 fixing plants and non-fixing plants are simultaneously grown. This method, which compared N yields of fixing and non-fixing plants, is often used in the fields to obtain rough estimates of N_2 fixed by legumes.

N-Balance Method

If all the inputs of nitrogen to a given field through precipitation, fertilizer etc. (except N_2 fixation are measured and all removal of N from that field is also measured (leaching, denitrification and crop harvesting), the positive difference between input and output can be ascribed to BNF.

— • —

Chapter 5

Azotobacter

Azotobacter are aerobic free living, heterotrophic N_2 – fixers. Of the several species of *Azotobacter, A. chroococcum* is found to be the dominant inhabitant in arable soils capable of fixing N_2 (2-15 mg N_2 fixed/g of carbon source) in culture media. The numbers of A. chroococcum in Indian soils rarely exceeds 10^5/g soil due to lack of organic matter and the presence of antagonistic microorganisms in soil. The bacterium produces an anti-fungal antibiotic, which inhibits the growth of several pathogenic fungi in the root region thereby preventing seedling mortality to a certain extent. Seed germination and plant stand are improved in plants upon inoculation with improved stains of *A.chroococcum* capable of elaborating growth substance even though quantitative assessment of these physiological attributes in situ in soil has not been made. Besides, nitrogen fixation, *Azotobacter* also produces thiamin, riboflavin, Indole acetic acid, and gibberellins. When *Azotobacter* applied to the seeds, seed germination is improved to a considerable extent.

CLASSIFICATION

Family	:	*Azotobacteriaceae*
Number of species	:	Seven (A.beijerinckii, *A.chroococum*, A.paspali, A.vinelandii, *A.agilis* Now categorized under the genus *A.azomonas)*, *A.insignis, A.mactroytogenes.*

Morphology

1. Cell size : Large ovoid cells, size ranging from 2.0-7.0 x 1.0-2.5 μ
2. Cell character : Polymorphic
3. Nitrogen source : N_2 through fixation, amino acids, N_2, NH_4^+, NO_3^-
4. Respiration : Aerobic.
5. Growth in media : Ashby/Jensen's medium
6. Doubling time : 3 hours

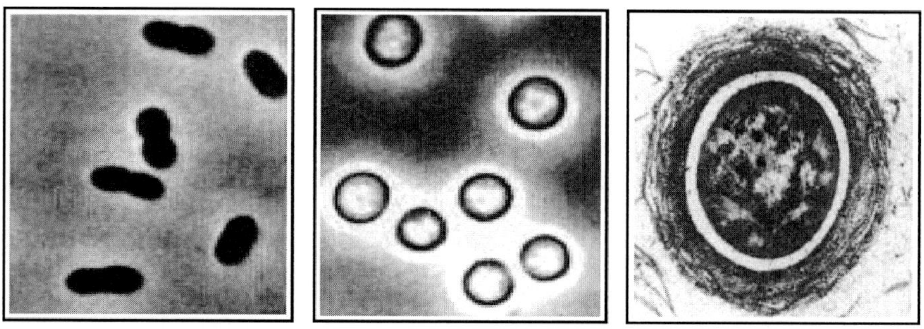

Azotobacter Cells and Cyst

Table 5.1. Colour Characteristics of *Azotobacter species*

Azotobacter Species	Colour Produced		
	Insoluble	**Soluble**	**Fluorescent**
A.chroococum	***Brown***	-	-
A.beijerinkia	Light Brown to sepia	-	-
A.vinelandi	-	Green	Apple green
A.macrocytogens	-	Purple	-
A.agilis	-	Green	Bluish White
A.insignis	-	Green	-

Contribution

1. 20-40 mg BNF/g of C-source in laboratory condition equivalent to 20-40 kg N/ha.

2. Production of growth promoting substances like vitamins of B-groups, Indole acetic Acid and gibberellic acid.
3. Biological control of soil borne plant disease by suppressing *Aspergillus, Fusarium.*

Recommended Crops

Rice, wheat, millets, other cereals, cotton, vegetables, sunflower, mustard, flowers. etc.

ISOLATION

Isolation of the free-living nitrogen fixing bacteria-*Azotobacter* from the rhizosphere soil of cereal crops.

MATERIALS

i. Rhizosphere soil samples (freshly collected)
ii. Sterile water blanks
iii. Sterile pipettes, Petridishes etc.
iv. Ashby's Mannitol agar medium

PROCEDURE

- Prepare Ashby's mannitol agar medium, sterilize and keep ready.
- Collect rhizosphere soil samples from any young cereal or vegetable crops
- Prepare serial dilutions of the soil sample up to 10^{-2}
- Transfer 1 ml each from 10^{-1} and ^{10} $^{-2}$ dilutions to each of three petridishes separately with the help of sterile pipette.
- Pour the sterilized melted and appropriately cooled Ashby's mannitol agar medium to the Petridishes and rotate in clockwise and anticlockwise direction gently so that the soil suspension distributed uniformly.
- Incubate the plates at room temperature for 4-5 days.

OBSERVATIONS

Observe the plates after 4-5 days for growth of *Azotobacter. Azotobacter* cells are often quite large showing slimy growth on the agar surface.

Aged cells are known to produce pigments and the pigmentation may be brown/yellow/grayish/green fluorescent/black depending on the species.

ENUMERATION OF AZOTOBACTER FROM THE SOIL

Aim

To enumerate the population of *Azotobacter* from the rhizosphere / non-rhizosphere soil.

MATERIALS

1. Soil samples
2. Ashby's mannitol agar medium
3. Sterile water blanks
4. Sterile Petridishes, pipettes etc.

PROCEDURE

- Prepare and sterilize the Ashby's mannitol agar.
- Add about 10.0 g soil to 100 ml sterile water blank, shake vigorously and allow to settle.
- Transfer 10 ml from the soil suspension (10^{-1}) to 90 ml water blank and prepare serial dilution up to 10^{-2}
- Pipette out 1 ml of suspension from 10^{-2} dilution to each of the three sterile Petridishes.
- Pour melted and cooled Ashby's mannitol agar medium to each of the three Petridishes (approximately 15-20 ml). Rotate the Petridishes gently in clockwise and anticlockwise direction for uniform spreading of the sample and allow the media to solidify.
- Incubate the plates under room temperature for 4-5 days.
- Find out the moisture content of soil sample by evaporating the soil suspension in the first dilution flask by keeping on a water bath and work out the dry weight of the soil sample taken.

OBSERVATIONS

- Observe the plates after incubation for the growth of *Azotobacter colonies*. Azotobacter colonies appear white, slimy on agar surface initially.

- Count the number of colonies appeared on the agar surface and calculate the average number of colonies of three plates.

CALCULATION

Calculate the number of colonies of Azotobacter per g of soil on dry weight basis as follows

$$\text{Population / g dry soil} = \frac{\text{Average no. of colonies X Dilution factor}}{\text{Dry weight of soil taken}}$$

—•—

Chapter 6

Azospirillum

The concept of "associative symbiosis" has widened our knowledge of free-living nitrogen fixing microorganisms. The bacterium was found to fix nitrogen in association with roots of various plants. This startling new discovery opened new vistas for world's most staple food crops like rice, maize, sorghum, wheat and millets, which were previously believed to have no potential for biological nitrogen fixation.

Azospirillum spp. are of ubiquitous distribution in many parts of the world and also with various standing crops grown in a variety of soil types. Lakshmi kumari et al (1976) first time reported the occurrence of nitrogen fixing spirillum in the roots of rice, sorghum and maize.. It is reported that various geographical regions of the world were rich in *Azospirillum*. It is also reported that nearly 10 per cent of the soils and roots from temperate region have Azospirillum while more than 50 per cent tropical soils were positive for *Azospirillum.*

In India, Rao et al. in 1978 reported the presence of *Azospirillum* in different rice soil types (alluvial, laterite, pokkali and kari). The occurrence of *Azospirillum* in the roots of cotton, maize, ragi, kudiraivali, guinea grass, hariyali grass, plantation crops, orchard plants, pearl millet, arecanut, banana, cashew, coconut, cocoa, rubber and cardamom and upland rice was well documented.

A new acid tolerant species viz., *Azospirillum amazonense* was reported from Amazon region. Charyulu and Rao in 1980 observed the occurrence of *Azospirillum* with appreciable nitrogen fixing ability in highly acidic soils with extremely low pH, Salt tolerant *Azospirillum* isolates have also been reported.

So far taxonomists have identified five species in the genus *Azospirillum* viz., *A.brasilense, A.lipoferum, A.amazonese, A.halopraeferens* and *A.irakense*.. They are Gram negative, vibrioid in shape, motile with a long polar flagellum and occasionally with peritrichous flagella. Organic acids such as malate and succinate are the preferred carbon sources for *Azospirillum*. However the different species vary in their ability to use different carbon sources. Among the free-living nitrogen-fixing bacteria *Azospirill*um is considered to be more efficient with nitrogenase properties comparatively better than other nitrogen fixers. It has been well demonstrated that *Azospirillum* inoculated plants were able to absorb nutrients from solutions at faster rates than uninoculated plants resulting in accumulation of more dry matter, N, P and K in the stem and leaves.

CLASSIFICATION

1. Family : *Spirillaceae*
2. Number of species : *(A.llipoferum, A.brasilense, A.amazonese, A.halopraeferns/seropedicae)*

Morphology

1. Cell size : Curved rods, 1.0 mm in diameter, size and shape vary
2. Reserve Material : Poly-B-hydroxy butyric acid
3. Gram reaction : Gram negative

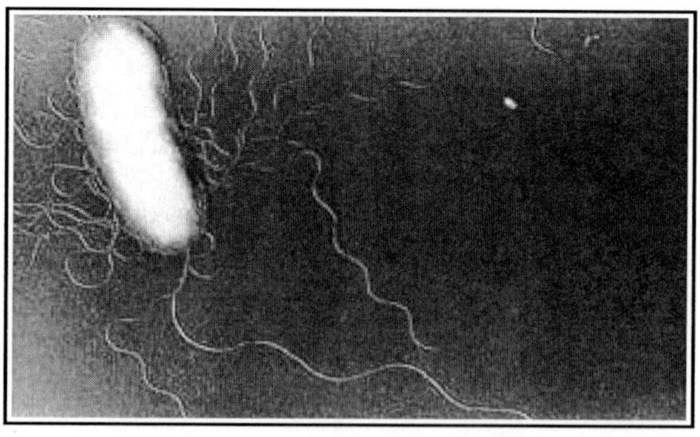

Electron micrograph of *Azospirillum* showing flagella

Physiology

1. Nature : Chemoheterotrophic, associative.
2. Sole carbon source : Organic acids, L-arabinose, D-gluconate, D-fructose, D-glucose, sucrose, pectin.
3. Nitrogen source : N_2 through fixation, amino acids, N_2, NH_4^+, NO_3^-
4. Respiration : Aerobic, Microaerophilic .
5. Growth in media : Nitrogen free Bromothymol Blue (NFB) medium
6. Doubling time : 1.0 hour on ammonia containing medium, 5.5 to 7 hour on a malete containing semi-solid medium.

Table 6.1. Characteristics of *Azospirillum* Species

Medium	***A .lipoferum***	***A. brasilense***
Semi-solid NFB medium with Glucose or keto glutaric acid as sole carbon source incubated at 37°C	Heavy growth on surface after 72 hours	No growth or fine pellicle much below surface after 72 hours
Semi-solid NFB medium with 0.005% Yeast extract	No growth after several successive transfers	Normal growth.
Cell in 48-hold cultures in semi solid NFB medium with 0.005% yeast extract.	Plemorphic, larger shaped or helical shaped (1,4-1, 7 x 5.30 µm), slow or non-motile with prominent PHB granules.	Short, slightly curved, very motile rods (1.0 x 2.1 - 3.8µm) with prominent PHB granules. Retain their cell form and characteristic motility for many days.

Contribution

1. 10-20 mg N fixed/ g malate under laboratory condition equivalent to 20-40 kg N ha^{-1}.
2. Results in increased mineral and water uptake, root development, vegetative growth and crop yield 15-30%.

Recommended for: Rice, Millets maize, wheat, sorghum, sugarcane, and co-inoculant for legumes.

Response: Average increase in yield 15-30%.

ISOLATION

Isolation of associative symbiotic nitrogen fixing bacterium-Azospirillum from the roots of cereal crops.

Materials

1. Freshly collected plant roots
2. Sterile water blanks
3. N-free malic acid semisolid medium
4. Sterile Petridishes
5. Forceps etc.

Procedure

- Prepare N-free malic acid semi solid medium, dispense in 5 ml quantities in the test tubes. Sterilize and keep ready.
- Collect fresh roots/rhizosphere soil of cereal crops from the experimental field.
- Mix known quality of the root/soil with sterile water blank and prepare serial dilution upto $10^{-6.}$
- Pipette out aseptically 1 ml from10^{-4} dilution to each of the 5 tubes containing malic acid semisolid medium. Similarly transfer 1 ml from 10^{-5} to each of 5 tubes and 1 ml from 10^{-6} dilution to each of the 5 tubes of medium.
- Incubate the inoculated tubes at root temperature of 30±1°C for a period of one week.
- Maintain control tubes without adding the root sample.

Observations

Observe the inoculated tubes for the change in colour from yellowish green to brilliant blue and for the formation of sub surface pellicles.

Calculation

Count the number of tubes showing positive response (as above) in each dilution.

Find out the MPN index using statistical most probable number table.

Determine the moisture content of the sample and work out the dry weight of sample taken.

Multiply the MPN index with the middle dilution used (10^{-5}) and calculate the population per g dry weight of sample

i.e.

$$\frac{\text{MPN index} \times 10^{5}}{\text{Dry wt of sample}} = \text{Population/g sample on dry wt. basis}$$

PURITY CHECK FOR *AZOSPIRILLUM* CULTURES

To check the purity of *Azospirillum* culture using potato infusion agar.

Materials

1. *Azospirillum* cultures in master plate/slant
2. Potato infusion agar
3. Petridishes etc

Procedure

- Prepare the potato infusion agar medium as per composition and sterilize.
- Prepare potato infusion agar plates by pouring the medium (20 ml/plate) and allow to solidify.
- Streak the Azospirillum colonies from master plate/slant on the potato infusion agar surface.
- Incubate the plates at 30 ± 1°C for 5-7 days.

Observations

Observe the plates for *Azospirillum* growth after incubation period. *Azospirillum* colonies are characteristically more or less pinkish, usually wrinkled and partially growing into the agar.

CHARACTERISTICS OF *AZOSPIRILLUM*

To recognize *Azospirillum* cultures the following criteria are employed:

- Change of the colour of the growth medium (normally N-free malic acid semi solid or broth medium) from yellowish green to brilliant blue.
- Formation of typical subsurface white granular pellicles
- Fast moving spiral shaped large cells when observed under light microscope using wet mounts.
- Typically curved cells with poly-B hydroxyl butyrate granules when observed under oil immersion microscope.
- Characteristically more or less pinkish usually wrinkled colonies, partially growing into the agar in the potato infusion agar plates.

TEST FOR EFFECTIVE ROOT GROWTH

Pot Culture Test

After three weeks of growth thin down the number of plants in each pot to four uniform plants. At the end of three months, take one set of pot from control and inoculated series and separate the plants carefully from the soil under slow running water. Obtain the number, length and mass of the plant as a whole including branch, seeds if any and roots. At the end of three months, harvest the shoot system (depending on the crop). Dry at 60°C for 48 hours and determine the dry mass. For the above purpose maintain a minimum of four replications or a maximum of 16 replications.

Better vigour of roots in inoculated seeds if established from the data when compared to the control, is the confirmation of effectiveness of biofertilizer. If there is 10 percent increase in the plants over the dry mass of uninoculated control without nitrate, it may be concluded that the culture is of required quality.

METHOD FOR ESTIMATING MPN COUNT FOR *AZOSPIRILLUM*

(a) Add 100g-biofertilizer samples to 900 ml of sterile distilled water.

(b) Shake for 10 minutes on a reciprocal shaker.

(c) Make ten fold dilution series.

(d) Pipette 1 ml of each dilution (from 10^{-1} to 10^{-7}) to each one of 5 replicates (5 tubes containing nitrogen free bromothymol blue semi-solid malate media for *Azospirillum*).

(e) Begin by taking aliquot from the highest dilution and proceed down the series with same pipette.

(f) Collect the tubes from 10^5, 10^6 and 10^7 dilution level.

(g) Incubate the inoculated series for 2 days at 30°C.

(h) Check the characteristic sub-surface white pellicles (+ or -) in malate semi-solid medium, a change to dark blue colour of the medium.

Calculation

a. To calculate the most probable number of organisms in the original sample, select as P_1 the number of positive tubes in the least concentrated dilution in which all tubes are a positive or in which the greatest number of tubes is + ve, and let P_2 and P_3 represent the number of positive tubes in the next two higher dilution.

b. Then find the row of numbers in Table in which P_1 and P_2 corresponded to the values observed experimentally. Follow that row of numbers across the table to the column headed by the observed value of P.

c. The figure at the point of intersection is the most probable number of organisms in the quality of original sample represented in the inoculum added in the second dilution. Multiply this figure by the appropriate dilution factor to obtain the MPN value.

Table 6.2. Most Probable Numbers for use with 10 Fold dilutions and 5 Tubes per Dilution

		Most probable number for Indicated values of P_2					
P_1	P_2	0	1	2	3	4	5
0	0	0.018	0.036	0.054	0.072	0.090	-
0	1	0.018	0.0.36	0.055	0.073	0.091	0.11
0	2	.0037	0.055	0.074	0.092	0.11	0.13
0	3	0.056	0.074	0.093	0.11	0.13	0.15

Contd...

Table 6.2. Contd...

0	4	0.075	0.094	0.11	0.13	0.15	0.17
0	5	0.094	0.11	0.13	0.15	0.17	0.19
1	0	0.020	0.04	0.06	0.08	0.10	0.12
1	1	0.040	0.061	0.081	0.10	0.12	0.14
1	2	0.061	0.082	0.10	0.12	0.16	0.17
1	3	0.089	0.10	0.13	0.16	0.17	0.19
1	4	0.11	0.13	0.15	0.17	0.19	0.22
1	5	0.13	0.15	0.17	0.19	0.22	0.24
2	0	0.046	0.068	0.091	0.12	0.14	0.16
2	1	0.068	0.092	0.12	0.14	0.17	0.19
2	2	0.093	0.12	0.14	0.17	0.19	0.22
2	3	0.12	0.14	0.17	0.20	0.22	0.25
2	4	0.15	0.17	0.20	0.23	0.25	0.28
2	5	0.17	0.20	0.23	0.26	0.29	0.34
3	3	0.078	0.11	0.13	0.16	0.20	0.23
3	1	0.11	0.14	0.17	0.20	0.23	0.27
3	2	0.14	0.17	0.20	0.24	0.27	0.31
3	3	0.17	0.21	0.24	0.28	0.31	0.35
3	4	0.21	0.24	0.28	0.32	0.36	0.40
3	5	0.25	0.29	0.32	0.37	0.41	0.45
4	0	0.13	0.17	0.21	0.25	0.30	0.36
4	1	0.17	0.21	0.26	0.31	0.36	0.42
4	2	0.22	0.26	0.32	0.38	0.44	0.50
4	3	0.27	0.33	0.39	0.45	0.52	0.59
4	4	0.34	0.40	0.47	0.54	0.62	0.69
4	5	0.41	0.48	0.56	0.64	0.72	0.81
5	0	0.23	0.31	0.43	0.58	0.76	0.95
5	1	0.33	0.46	0.64	0.84	1.1	1.3
5	2	0.49	0.70	0.95	1.2	1.5	1.8
5	3	0.79	1.1	1.4	1.8	2.1	2.5
5	4	1.3	1.7	2.2	2.8	3.5	4.3
5	5	2.4	3.5	5.4	9.2	16	-

—•—

Chapter 7

Acetobacter

Acetobacter is an obligatory aerobic, nitrogen-fixing bacteria that is known for producing acid as a result of metabolic processes. While all nitrogen-fixing bacteria contain nitrogenase in order to utilize atmospheric nitrogen gas as a source for metabolic biosynthesis, different nitrogen-fixing microorganisms protect the oxygen-sensitive microorganisms from oxygen exposure in different ways. *A. diazotrophicus* has been called interesting because it carries out nitrogen fixation under aerobic conditions. It needs oxygen for the production of large quantities of ATP required for nitrogen fixation; however, little is known about the mechanism or system that protects the nitrogenase under aerobic conditions. *A. diazotrophicus* is a plant endophyte and has been said to be capable of excreting about half of its fixed nitrogen in a form that plants can use.

DESCRIPTION

Acetobacter bacteria, such as *Acetobacter diazotrophicus* that can be isolated from coffee plants or sugarcane, are acid producing, nitrogen-fixing bacteria. In fact, the *A. diazotrophicus*-sugarcane relationship, first observed in Brazil, was the first report of a beneficial symbiotic relationship between grasses and bacteria through nitrogen fixation. Nitrogen-fixing bacteria are important in modern agriculture and exploiting these bacteria would decrease the present dependency on nitrogen fertilizers, which would have positive results for the ecosystem and the health of humans and other animals.

CLASSIFICATION

1. Family : *Acetobacteraceae*
2. Number of species : *A. diazotrophicus*

Morphology

1. Cell size : Ellipsoidal rod shaped cells
2. Gram reaction : Gram negative

They can grow on ethanol, methanol, acetate, amino acids and produce H_2S. They oxidize acetate and lactate to CO_2 and water. 10 % sucrose solution supports growth.

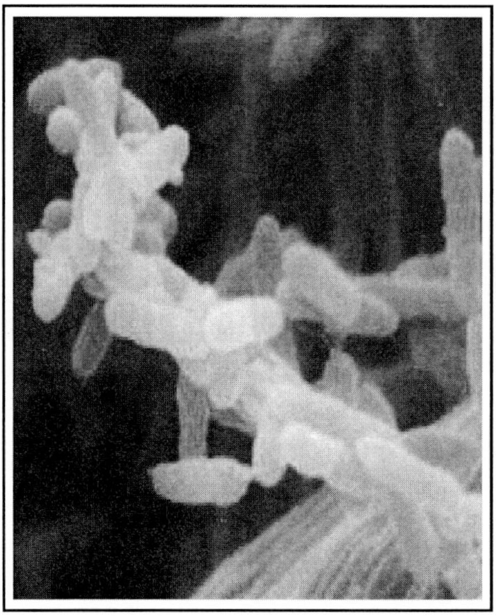

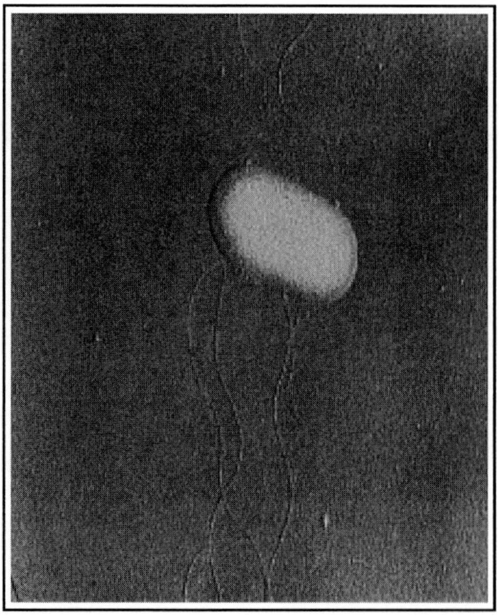

Acetobacter Diazotrophicus

ECOLOGY

Acetobacter bateria can be found in symbiotic relationships with many different plants, such as sugarcane and coffee plants, as well as in fermenting vinegar. Endophytes are prokaryotes that associate with plants by colonizing their internal tissues. Many of these bacteria have been observed to promote plant growth, but the mechanisms behind this relationship are not yet fully understood. Below is a table of plant part where *A. diazotrophicus* can be found.

Plant	Parts
Sugarcane	root, root hair, stem, leaf
Cameroon grass	root, stem
Sweet potato	root, stem, tuber
Coffee	root, rhizosphere, stem
Ragi	root, rhizosphere, stem
Tea	root
Pineapple	fruit
Mango	fruit
Banana	rhizosphere
Others : mealy bugs, VAM spores	internal environment

Contribution

1. Can fix up to 100 kg N /ha in sugarcane and increase in cane yield by 8 - 10 %.
2. Plant growth acceleration due to growth hormones.

Isolation

Isolation of an endophyte *Acetobacter* from the roots, stem of sugarcane.

Materials

1. Freshly collected plant roots/stem
2. Sterile water blanks
3. LGIP medium
4. Sterile Petri dishes
5. Sterile Petri dishes
6. Forceps etc.

Procedure

- Prepare LGIP medium, and pour in Petri dishes.
- Collect fresh roots/stem of sugarcane from the experimental field.

- Mix known quality of the root/soil with sterile water blank and prepare serial dilution up to 10^{-6}.
- Pipette out aseptically 0.01 ml from $10^{-4,}$ 10^{-5} and 10^{-6} dilution to Petri dishes containing LGIP medium. Incubate the inoculated tubes at root temperature of 28±1°C for a period of one weed.

Observations

Observe the inoculated Petri dishes for orange colored colonies.

—•—

Chapter 8

Phosphate Solubilizing Bacteria

Phosphate solubilizing bacteria otherwise called phosphobacteria play a major role in the solubilization and uptake of native and applied soil phosphorus requirement of crops. Efficiency of applied phosphorus is very low due to its fixation either in the form of aluminum or iron phosphates in acidic soils or in the form of calcium phosphate in neutral or alkaline soils. The introduction of efficient 'P' solubilizers in the rhizosphere of crops and in soils increases the availability of phosphorus. Phosphobacteria bring about dissolution of bound forms of phosphates in soil and hence advocated as biofertilizer to crop plants. Several soil bacteria and fungi notably species of *Bacillus, Pseudomonas, Asperigillus, Penicillium* secrete organic acids and lower the pH in their vicinity to bring about solubilisation of bound phosphates in soils.

Beneficial effects due to inoculation with phosphate solubilizing organisms had been reported for different crops under diverse agro climatic conditions. There is a report of increase in yield accompanied with high phosphorus uptake in plants inoculated with soil containing phosphate-solubilizing microorganisms (PSM). Phosphobateria enhanced the efficiency of plant uptake of phosphatic fertilizer particularly super phosphate in maize. Inoculation of composite culture of PSM comprising of *Pseudomonas striata* and *Aspergillus awarm*ori recorded increase in the yield of wheat and potato.

Table 8.1. Classification

Microorganism	Important Species
Bacteria	*i) Bacillus megaterium var, phosphaticum, B.circulans, B.subtilis, B.polymyxa*
	ii) B. coagulans
	iii) Pseudomonas striata, P.liquifacines; P.putida
Fungi	*i) Aspergillus awamori, A.fumigatus, A.flavus,*
	ii) Penicillium digitatum, P.lilacinum,
Yeast	*iii) Torulospora globosa*

(Some *Streptomyces sp.* can also solubilise P. significantly).

Table 8.2. Morphology and physiology

Morphology			
Micro-Organism	**Cell Size**	**Cell Character**	**Gram Reaction**
Bacteria	1.1 to 2.2 µm, rod shaped	Transparent zones of clearing around microbial colonies indicate extent of P solubilisation, spreading type growth, and spore formation for Bacillus.	*Bacillus* Gram + ve
Fungi	For *Asperigillus*, conidial heads coloured, conidiophores pitted, often appearing rough in general, arise separately from the substratum, 400-700µ or 1000µ long x 5-15µ in diameter, For *Penicillium*, typically in single verticles of phialides borne or branches or consisting of three more series of verticles.		
Physiology:			
Bacteria	Aerobic, hetertrophic. Glucose is the main carbon source but they can utilize other carbon source also.		
Fungi	Most molds are strictly aerobic. Glucose is a suitable carbon source. other sugars can also be utilized including complex carbon compound.		

All these organisms can be grown in Pikovskaya's medium.

Contribution

- These microorganisms possess the ability to bring insoluble soil phosphate into soluble forms by secreting several organic acids.
- Under favourable conditions, they can solubilise 20-30% of insoluble phosphate and yield may increase by 10 to 20%.

ISOLATION AND ENUMERATION OF PHOSPHOBACTERIA FROM SOIL

Isolation and enumeration of phophobacteria from soil.

Materials

1. Sterile Water blanks - 100 ml and 9 ml.
2. Soil/Rhizosphere soil sample.
3. Sterile Petri plates and pipettes.
4. Pikovskaya's Medium.

Procedure

- Prepare soil or rhizosphere soil suspensions in sterile water blanks taking 10 g soil sample and 1 g rhizosphere sample respectively.
- Prepare serial dilutions of the soil or rhizosphere soil suspensions.
- Pipette out 1 ml into each petri plate from the 10^{-5} dilution in the case of soil sample and 10^{-6} dilution in the case of rhizosphere samples. Keep 4 plates for each treatment.
- Pour about 10-15 ml of the above medium into each petri-plate rotate well to distribute the medium and incubate at 25-28°C.

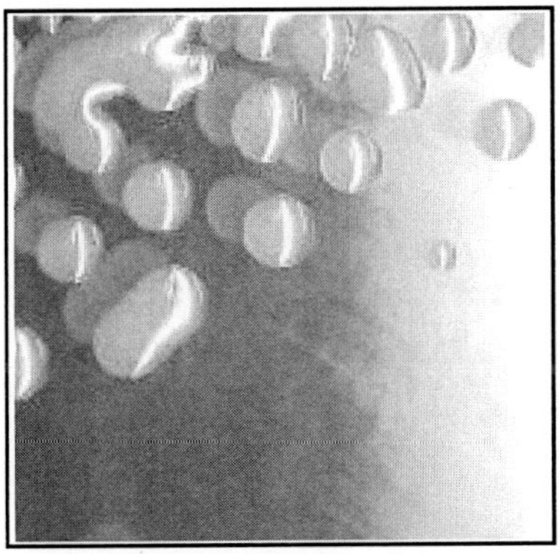

Mucoid Colony of PSM - *Bacillus* sp.

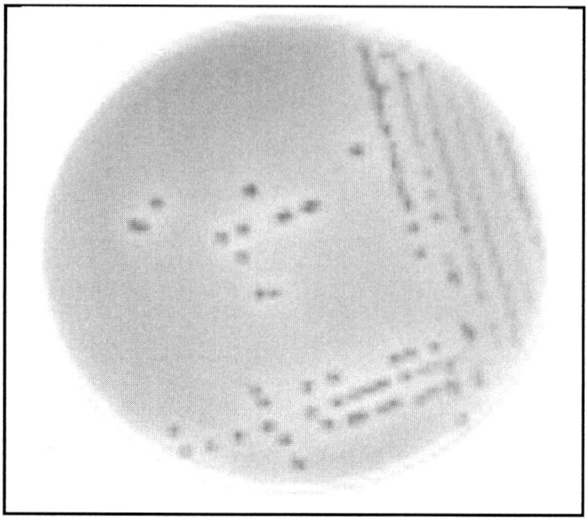

Phosphate solubilization Zone by PSM

Observations

Observe the clear "halo zone" around bacterial colonies surrounded by a turbid white background after 3-5 days of incubation. Count the number of "halos" in each plate and arrive at the average number of phosphate dissolving bacteria in each case. Calculate the population of phophobacteria per gram of soil or rhizosphere sample and express on dry weight basis.

— • —

Chapter 9

Mass Production of Bacterial Biofertilizers

Production of bacterial biofertilizers viz., *Rhizobium, Azospirillum, Azotobacter, Acetobacte*r and Phosphobacteria in large scale.

MATERIALS

1. Inoculum of *Rhizobium, Azospirillum, Azotobacter, Acetobacter* and Phosphobacteria at logarithmic growth phase. (Active young culture)
2. Carrier materials
3. Polythene bags
4. Balance
5. Trays
6. Balance
7. Gloves
8. Sterile beaker of 200 ml capacity
9. Electric plastic bag sealer etc.

PROCEDURE

Inoculum Preparation

- Prepare the respective broth media for *Rhizobium, Azospirillum, Azotobacter Acetobacter* and Phosphobacteria in 250 ml, 3 lit. and 5 lit, flask and sterilize.

- Inoculate the media in 250 ml flask with efficient strains of the concerned organism under aseptic conditions.
- Incubate the flasks under shaking condition in gyratory shakers at room temperature for 2 days (Phosphobacteria). 3-5 days (*Rhizobium*) and 5-7 days (*Azotobacter, Azospirillum,* and Acetobacter).
- Check the flask for growth and determine the population load. This will serve as the STARTER CULTURE.
- Using this starter culture (at log phase) inoculate the large flask (3 lit. & 5 lit.) containing media for the respective organisms.
- Incubate the large flasks at room temperature with occasional shaking for the growth period mentioned above.
- Prepare the concerned media in large quantities in 'FERMENTOR', sterilize well, cool and keep ready.
- Inoculate the media in fermenter with the log phase culture grown in large flasks (usually 1-2 % inoculum is sufficient, however inoculation may be done up to 5 % depending on the growth).
- Prior to inoculation determine the population load of the organism in the broth used for inoculation.
- Grow the cells in fermenter by providing aeration (passing sterile air through compressor and sterilizing agents like glass wool, cotton wool, acid, etc.,) and steaming.
- Check the population of inoculated organism and contamination if any at growth period.
- Harvest the cell (Spent broth) after incubation period (as mentioned above) and use for inoculant preparation.
- Determine the population load at harvest and check bacterial/ fungal contamination in the spent broth.

Note

The population of the organism should be more than 10^9 cells per ml of the fermented broth.

There should not be any fungal or any other bacterial contamination at 10^{-6} dilution level.

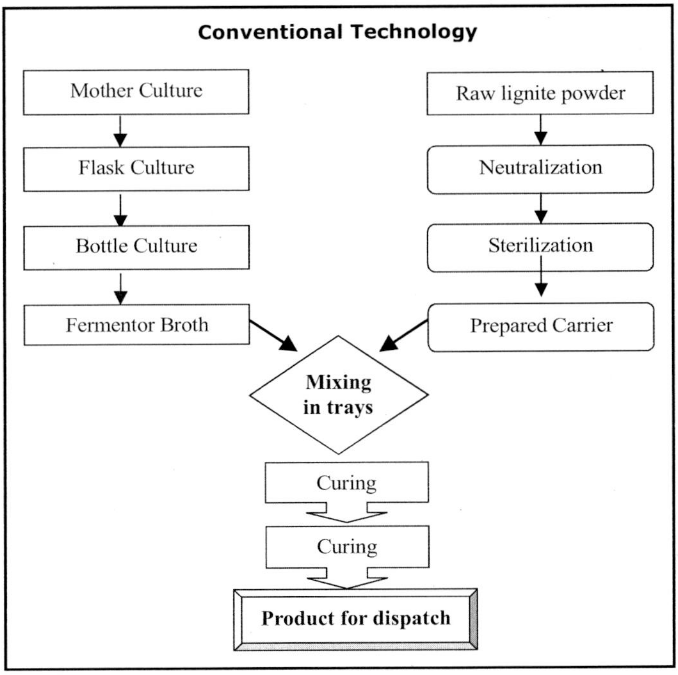

Flow chart showing conventional Biofertiliser Production Process

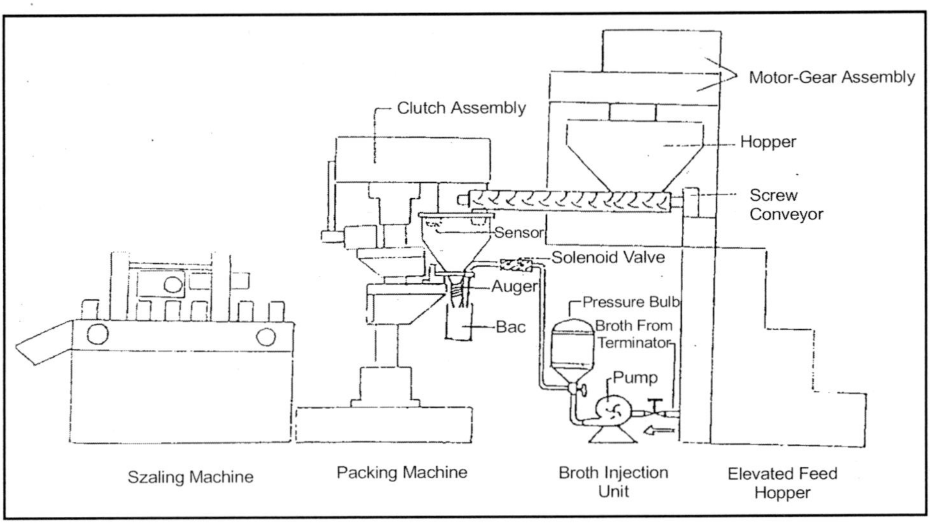

Semi-automatic packing system for biofertilizers

It is not advisable to store the broth after fermentation for periods longer than 24 hours. Even at 4°C number of viable cells begin to decrease.

Preparation of Carrier Materials

- Powder the carrier material (peat soil or lignite) to a fine powder so as to pass through 212 micron IS sieve.
- Neutralize the pH of the carrier material with the help of calcium carbonate powder (since peat soil & lignite are acidic in nature).
- Sterilize the neutralized carrier material in an autoclave to eliminate the contaminants.

Preparation of Inoculant Packet

- Spread the neutralized sterilized carrier material in clean dry sterile metallic trays.
- Pour the appropriately grown bacterial (*Rhizobium, Azospirillum, Azotobacter, Acetobacter,* Phosphobacteria) cultures on the carrier and mix well manually (by wearing sterile gloves) or by mechanical mixer (The culture suspension is to be added to a level of 40-50 % water holding capacity depending upon the population).
- Pack the inoculant in 200 g quantities in polythene bags. Seal with electric sealer and allow for curing for 2-3 days at room temperature (Curing can also be done by spreading the inoculants on a clean floor/polythene sheet/by keeping in open shallow tubs/trays with polythene covering for 2-3 days at room temperature before packeting).
- Check the initial population load of organism after curing in the inoculant packet and record.

Specification of Polythene Bag

- The polythene bags should be of low-density grade.
- The thickness of the bag should be around 50-75 micron.
- Each packet should be marked with the name of the product, strain number, the crop to which recommended, method of inoculation, date of manufacture, batch no. date of expiry, price, full address of the manufacturer and storage instructions etc.

Storage of Inoculant Packets

- Store the inoculant packets in a cool place away from heat or direct sunlight.

- Store packets at room temperature or in cold storage conditions in lots in plastic crates or polythene/gunny bags.
- Determine the population of the organism in the inoculant packet at 15 days interval. The population should be more than 10^8 cells/g for *Rhizobium,* 10^9 cells/g for *Azospirillum* and 10^7 cells/g of the carrier material for *Azotobacter* on dry weight basis immediately after curing and should not be less than 10^7 cells/g for *Rhizobium* and *Azospirillum* and 10^6 cells/g for *Azotobacter* carrier material on dry weight basis 15 days before the expiry date.

TYPES OF RAW MATERIALS

Carrier

Carrier is the most important ingredient in biofertilizer formulation. It carries the live microorganisms, which function as biofertilizer. The shelf life of a biofertilizer also depends on the quality of the carrier. As per ISI specification, the carrier should be in powder form and capable of passing through 150-212 micron (72-100 mesh) IS sieve. The qualities of a good carrier material are:

- Highly absorptive (water holding capacity) and easy to process.
- Non-toxic to microorganisms.
- Easy to sterilize (by autoclaving or gamma-irradiation).
- Available in adequate amounts and inexpensive.
- Provide good adhesion to seed.
- Have pH buffering capacity (high cation and/or anion exchange capacity).

A wide range of carriers have been developed for carrying microbial inoculants in acceptable high numbers. These include peat, lignite, charcoal, rice husk, vermiculite, soil, coir dust, press mud, polymer compound etc. In Australia, France and USA, where Rhizobium industry has become successful, peat is used as a carrier. Peat is dark brown or black residue produced by the partial decomposition and disintegrations of mosses, trees and other plants.

In India, peat contains only 54 % organic carbon as compared to Australian peat (65 %) and American peat (86 %). Indian peat is thus not of good quality. It is available only in the Nilgiri valley, (Tamil Nadu) and involves high transport cost in addition to quality variation.

The same is also applicable for lignite. Lignite is brownish black coal in which the alteration of vegetal material has proceeded further than in peat but not so far as in sub-bituminous coal. Charcoal is available in plenty but is costlier than peat and lignite. Besides, the quality of charcoal is always questionable. Among the properties which characterize charcoal, the more significant are: weight of the charcoal, content of volatiles, fixed carbon content, ash content, hardness and active surface.

It has been experienced that some biofertilizer production units use poor quality charcoal (result of incomplete hydrolysis), which has low water holding capacity, bulk density and porosity. Only the use of good quality charcoal can substitute peat or lignite as carrier. Details about peat, lignite and charcoal have been summarized in Table. Some manufactures in India and elsewhere are using vermiculite as carrier. It has been claimed to be a better carrier than peat.

PACKING MATERIALS

The first biofertiliser (*Rhizobium* inoculants) was prepared in glass bottles because these withstood the required autoclaving, could be sealed with stoppers or threaded caps and allowed visual inspection by the prospective buyer. Then metal cans replaced glass bottles but these did not become popular. Bulk of the world's biofertiliser production is now marketed in plastic pouches.

The reasons for choosing plastic pouches are: 1. Ease in handling; 2. A high gas exchange; 3. Relatively low moisture transmission; and 4. Heat-sealable.

The factors to be kept in mind while selecting polythene pouches are:

- Effect of gamma irradiation
- Presence of entrapped air and gas exchange capacity
- Moisture content
- Autoclavable tolerance

Table 9.1. Chemical and physical characteristic of common carriers

Parameters	**Peat (Indian)**	**Lignite**	**Charcoal**
CHEMICAL			
1. Appearance	Dark brown	Brownish black	Black
2. pH	Acidic, needs neutralization	Acidic, needs neutralization	Neutral
3. Odour	None	None	None
4. Carbon (%)	54.20	72.79	84.00
5. Hydrogen (%)	5.70	4.74	2.30
6. Oxygen (%)	31.60	19.60	10.70
7. Nitrogen (%)	1.50	0.98	0.01
8. Ash (%)	7.00	7.00	3.00
PHYSICAL			
1. Water holding capacity (%)	149.30	198.90	200.00
2. Moisture (%)	10-50	40-50	20-30
3. Particle density (%)	2.18	1.66	1.62
4. Porosity (%)	53.16	34.79	73.39
5. Total surface area (m^2/g)	646.60	556.40	870.90
AVAILABILITY	5.5 million ton Nilgiri Valley (Tamil Nadu)	3 million ton a. Neyveli (TN) b. Singrauli (MP) c. Palana (Rajasthan)	Plenty Everywhere

The low-density polythene (LDPE) packets (0.038-0.051 mm) cannot be sterilized in an autoclave, as they cannot tolerate the 121°C temperature. The melting point of polythene rises with its increasing density. Therefore, high-density polythene packets (HDPE) (0.075-0.090 mm) may be used. For autoclaving, polythene packets are the best, having can tolerance at 121°C.

Table 9.2. Properties of Packaging Materials used for Biofertilizer

Property	LDPE	HDPE	Polypropylene
Molecular Weight	1500-7000	25000-50000	NA
Thickness (mm)	0.038-0.08	0.08-0.1	0.05-0.09
Appearance	Translucent	Translucent to Opaque	Transparent
Melting Point (°C)	80-90	105-120	120-130
Autoclavability (during sterilization)	Poor	Moderate	Good
Stability towards irradiation	Good	Good	Fair
Permeability to gases[CC's (STP) Cm^2/sec/cm Hg at 25°C x 10^{-8}]			
CO_2	28	6.5	6
N_2	7.5	1.6	2.0
O_2	2.5	0.5	0.4
Gas exchange property	Good	Moderate	Moderate

If gamma irradiation facilities are available, the use of LDPE is recommended as gas exchange is better in LDPE and it can tolerate gamma-irradiation, which cannot be tolerated by polypropylene. Previously, there was a practice of punching small holes in the bags but these days it is not recommended as it also allows entry of contaminants and the moisture loss is also high. In the sterile system of production, the hole is closed by sterile adhesive tape immediately after injecting the broth.

Adhesives

The use of a sticker to glue the inoculant onto the seed is a general recommendation for postproduction field application. A number of stickers such as wall paper glue (5%), gum Arabic (4%), carboxy methyl cellulose (4%), honey (10%), powdered milk (10%), gur or sugar (10%) etc are available and serve not only as glue for the bacteria to stick on to the seed but also serve as food for the bacteria until the infection process is complete in case of rhizobia.

The inclusion of adhesives within the prepared inoculant is desirable and practical but requires careful selection, as some adhesives would merely set within the packet after impregnation. In India, generally gum Arabic or gum acacia powder is used as an adhesive. These gums are viscous exudates from Acacia (babul) trees. Acacia has number of species like *A. nelotica* (Babul), *A. auriculiformis* (Bengali Babul) etc. Gum Arabic belongs to *A. arabic* and sometimes to *A. senegal*.

Specifications for Raw Materials

- *Wood Charcoal:* 50-200 mesh, good quality, pH 6.5-7.5, 120-140% water holding capacity, moisture percentage, 40% (w/w), ash content 5-10%
- *Raw Lignite Powder:* 150-200 mesh, 35% moisture percentage, finely powdered, duly packed in HDPE bag of 40-50 kg each.
- *Gum Arabic:* 150-200 mesh dry powder.
- *HDPE Polythene Bag:* made of high quality, 1st grade milky white polythene tube having thickness of 300 gauge and sealed from one end. Individual bag printed on both sides (Flexo printing) as per the design and size. Bag sizes range from 8 x 16 cm to 30 x 40 cm to contain 40 to 2880 g inoculant respectively: In India the standard packet size 15 x 25 cm for 200 g biofertilizers.
- *Instafix Gummed plastic Tape:* 5 cm wide tape for fixing and sealing the corrugated cardboard boxes.
- *LDPE Plastic Strips:* For packing the cardboard boxes made of first grade LDPE 5 cm wide thick in 200 m rolls.
- *Iron Clips:* For packing plastic strip of 5 cm sizes.
- *Aluminium foil:* 0.03 mm thickness.

These are current specification and can change with the development of new materials and changing needs of the production and marketing system.

Sterilization

Sterilization is a frequently used process in biofertilizer work.

Seed Sterilization

Seed sterilization is generally done by covering the seeds with sodium hypochlorite (NaOCl) for 4 minutes or with concentrated sulphuric acid for 10-15 minutes as and when required. In both the cases, after the treatment, seeds should be rinsed at least 9 times with sterile water and them soaked in water for 6-15 hours as per requirement.

Nodule Sterilization

Nodules are to be immersed in 95% ethanol for 5-10 seconds and then in sterilent and left for 3-4 minutes. Sterilent can be 7-8 % solution of sodium or calcium hypochlorite or 3% hydrogen peroxide or 1% acidified mercuric chloride solution.

Sterilization of Glasswares

Generally, glasswares like pipette, Petridishes etc. are sterilized by dry heat sterilization. This involves exposure for 1- 2 hour at 160°C-170° C in the hot air oven (sterilizer).

Sterilization of Broth

The broth is generally sterilized by means of moist heat. For upto 1000 ml broth, exposure time of 20-30 minutes is sufficient, but for 2000 ml, it will be 40-60 minutes and for more than 5000 ml, a minimum of 60 minutes in the autoclave is suggested.

Sterilization of Carrier

Carriers are sometimes sterilized and sometimes not but it is better to sterilize them. This is especially important when the bacteria are to make most of their growth in the carrier rather than in the broth culture. If, other bacteria are present in the carrier, they will grow along with the desired ones such as *Rhizobium* or Azotobacter. If their numbers are nearly as great as that of the *Rhizobium* added, or if they grow faster than *Rhizobium / Azotobacter*, there may be more unwanted microorganisms than biofertilizer in the final product. It is necessary to minimize the number of unwanted bacteria.

Sterilization for packaged carriers usually involves either autoclaving or gamma-irradiation. For sterilization in an autoclave, following two systems are followed:

- To sterilize the carrier in autoclave at 121°C for 4 hours.
- To sterilize the carrier for 1 hour at 121°C, 2 to3 times on alternative days.

The reason for the second system is that the carrier cannot be sterilized fully in autoclave. A single 4 hours pass often initiates temperature tolerant spores of fungi and actionomycetes etc. To germinate. This can suppress the growth of N-fixing bacteria such as *Rhizobium, Azotobacter, Azospirillum* and *Acetobacter.*

Sterilization of Packaging Materials

Biofertilizers are generally supplied in non-sterilized LDPE packets and thus there is a chance of contamination. This can be avoided if the packets are pre-sterilized. But LDPE packets cannot tolerate 121°C temperature and hence, are not autoclavable. Their sterilization is only possible through gamma-irradiation. If the packaging materials are made of polypropylene or HDPE, these can be sterilized by autoclave.

The polypropylene tested in India is only satisfactory when subjected to autoclave pressures of a maximum of 10 psi (0.7 kg/cm^2). Even at such a high pressure, it is essential to expel all air before sealing the package. Alternatively, the package can be unsealed but closed with a clip to allow equilibration during sterilization and subsequently sealed.

Marking

Each polyethylene pack shall be marked legibly with the following information:

1. Name of the product,
2. Name and address of the manufacturer;
3. Crop(s) for which intended;
4. Type of the carrier used;
5. Batch number;
6. Date of manufacture;
7. Expiry date which shall not be less than 6 months from the date of manufacture;
8. Net mass in kg and area meant for;
9. Storage instructions worded as under;
   "STORE IN COOL PLACE AWAY FROM DIRECT SUN LIGHT AND HEAT"
10. Any other information required under the Standards of Weights and Measures (packaged Commodities) Rules, 1977.

Direction for use of biofertilizer shall be printed briefly on the packets. If possible a separate pamphlet may preferably be given with it.

BIS CERTIFICATION MARKING

The product may also be marked with the BIS Standard Mark.

The use of the Standard Mark is governed by the provisions of Bureau of Indian Standards Act, 1986 and Rules and Regulations made there under. The details of conditions under which the license for the use of Standard Mark may be granted to manufactures or producers or may be obtained from the Bureau of Indian Standards.

Storage

Biofertilizer shall be stored after the manufacture in a cool and dry place away from direct heat preferably at a temperature of 20°C and not exceeding 30°C. It shall also be the duty of the manufacturer to instruct the retailers and, in turn, the users about the precautions to be taken during storage.

— • —

Chapter 10

Quality Control of Biofertilizers

INTRODUCTION

Quality of inoculants manufactured is one of the important factors resulting in their success or failure and acceptance or rejection by the farmers. Basically quality means the presence of the right type of microorganisms in active form and desired number. There are BIS standards for *Rhizobium, Azotobacter, Azospirillum* and phosphate culture, however, there is no systematic quality certification system and monitoring mechanisms. Also adoption of BIS standards is not compulsory, so hardly few manufacturers prepare biofertilizers as per BIS standards. Recently, Ministry of Agriculture has issued an amendment in fertilizer control Act 1985 and now it is termed as Fertilizer (Control) Amendment Order, 2006. Under this amendment, four biofertilizers viz. *Rhizobium, Azotobacter, Azospirillum* and Phosphate solubilzing bacteria are included .

As the products are entirely being living microorganisms, the quality check up and certification batch-wise is highly essential. Each manufactures should have laboratory infrastructure with the following facilities:

1. Adequate microbiological laboratory and qualified microbiologists.
2. Sampling and testing at various stages of production, including the quality if raw materials.

3. Specify on the packets all the contents and cell counts. The source of mother culture and strain name should also be mentioned.
4. The unit should fix their quality certificate and batch number, pack the products in proper packing materials.
5. Store the final product in cooler places till they are sold to the farmers.
6. Ensure to have aseptic conditions, cleanliness and contamination free production lines and housing.
7. Preferably use automatic and closed systems.

It is necessary to confirm to certain control measures before and after bacterial inoculants. The quality control tests are performed mainly at following stages.

1. During mother culture stage
2. During carrier selection and standardization.
3. In broth stage
4. While mixing of broth with carrier.
5. In finished product- carrier based culture
6. During storage

A. Qualitative Tests

- *Contamination:* Check for the freedom from visible contaminants Observe the broth carefully for the change in colour, precipitation or growth of fungi.
- *pH :* the pH of the bacterial broth should be between 6.5 to 7.5. If the pH is below 6.0 or above 8.0, suspect presence of contaminants and reject. Check the pH with pH meter or pH indicator strip.
- *Agglutination:* Broth suspension should give definite agglutination with 1/100 diluted specific antiserum on a slide. Allow to react for 30 minutes. Observe the slide for agglutination under low power of the compound microscope.
- *Growth on glucose peptone agar:* Streak loopful of culture on a slope of this medium, incubate at 30°c, examine after 24 hr. Marked growth, often associated with change of pH, indicates

gross contamination. Some strains of rhizobia show slight growth on this medium, but this is without appreciable change in pH.

- Smear preparation and gram staining.

B. Quantitative Tests

1. Total Count

This is performed by using 10^9 diluted broth culture with the help of Petroff-Hausser bacterial counting chamber using high power of compound microscope. This gives the first assessment of the count and can also reveal gross contamination such as coccoid forms, spirals or chain cells of various microorganisms. The total count will include dead cells also. Hence, it should be used as a guide. Culture should show 10^8 to 10^9 cells/ml, which is likely to provide 10^7 cells/ml in final product.

2. Total Viable Count

Requirements

Yeast extract Mannitol agar (with Congo red) medium (Annexure I).

Procedure

Serially dilute the broth culture to obtain dilutions of the order of 10^6 to 10^9. Plate 0.2 ml aliquots of the dilutions on CRYEMA medium incubate the plates in inverted position at 28±2°C for 2-7 days, depending upon the species of *Rhizobium.*

Observations

The viable cell count of *Rhizobium* in final broth from shake culture of fermentors shall not be less than 10^8 to 10^9 cells/ml

Acceptance or rejection of broth cultures: Broth cultures showing pH 6.5 -7.5, gram negative cells having an at least 500 x 10^8 cells/ml viable rhizobia of specific type and which are free off significant contaminant are accepted for the manufacture of carrier-based inoculum.

3. Total viable counts of packet or liquid formulation sample

Procedure

Dilute 10 g or 10 ml contents of each sample packed to obtain dilutions of the order of 10^7 to 10^9. Proceed as mentioned in B(2).

Observations

The viable *Rhizobium* count should not be less than 10^7 cells/g or ml and should also nodulate the specific legume.

As per BIS specifications, certain tests are required to be conducted, like number of cells, colony character, reaction etc.

1. Inoculation should contain a minimum of 10^8 viable cells/g of the carrier on dry weight basis at the time of manufacture and 10^7 cells on expiry date marked on the packet.
2. No contaminations at 10^5 dilutions.
3. pH of inoculant should be 6.0-7.5
4. Carrier material should pass through 106 micron size sieve.
5. Packing low-density polythene bags of 50-75 μm.
6. Each packet should be marked with the details-Name of the product, leguminous crop for which intended, strain number, date of manufacture, date of expiry, method of application and storage instructions.

RHIZOBIUM INOCULANT (RI) ISI STANDARDS (IS 8268-1986)

1.	Viable Cell Count	RI shall contain 10^7 viable bacterial cells/g of the carrier material or 10^7 cell/ml of liquid material before expiry
2.	pH	6.5-7.5
3.	Moisture %age	35-40%
4.	Shelf Life	Minimum 6 months
5.	Carrier Material	The carrier material such as peat, lignite, peat-soil, humus shall be in the form of powder capable of passing through 150-212 micron IS sieve.
6.	Contamination	Nil at 10^5 dilution
7.	Efficiency character	*Rhizobium* inoculants shall show effective nodulation. If good effective pink nodulation is obtained in the inoculated species together with total absence or some times presence of stray nodules in the controls, it should be concluded that RI contains effective Rhizobia. The total

		dry mass of inoculated plants shall be significantly higher than that of the uninoculated controls and at least 50 % more than the controlled one.
8.	Marking on Packet	Each packet shall be marked legibly to give the following: a) Name of the product b) Name and address of manufacture c) Name of carrier d) Batch number e) Date of mfg. f) Date of expiry g) Net mass h) Crops for which intended i) Storage instructions.

AZOTOBACTER INOCULANT (AI)

ISI STANDARDS (IS 9138: 1979)

1.	Viable Cell Count	AI shall contain minimum 10^7 viable bacterial cells/g of the carrier material or 10^7 cell/ml of liquid material before expiry
2.	pH	6.5-7.0
3.	Moisture %age	35-40%
4.	Shelf Life	Minimum 6 months
5.	Carrier Material	The carrier material such as peat, lignite, peat-soil, humus shall be in the form of powder capable of passing through 150-212 micron IS sieve.
6.	Contamination	Nil at 10^5 dilution
7.	Efficiency character	The minimum amount of nitrogen fixed by *Azotobacter chroococcum* shall not less than 10 mg/g of sucrose utilized. Azotobacter inoculants should be effective for germination of seeds.
8.	Marking on Packet	Each packet shall be marked legibly to give the following:

		a. Name of the product b. Name and address of manufacture c. Name of carrier d. Batch number e. Date of mfg. f. Date of expiry g. Net mass h. Crops for which intended i. Storage instructions.

AZOSPIRILLUM INOCULANT (ASI)

ISI STANDARDS (IS 14806-2000)

1.	Viable Cell Count	ASI shall contain 10^7 viable bacterial cells/g of the carrier material or 10^7 cell/ml of liquid material before expiry
2.	pH	6.5 -7.5
3.	Moisture %age	35-40 %
4.	Shelf Life	Minimum 6 months
5.	Carrier Material	The carrier material such as peat, lignite, peat-soil, humus shall be in the form of powder capable of passing through 150-212 micron IS sieve.
6.	Contamination	Nil at 10^5 dilution
7.	Efficiency character	Formation of white pellicle in semisolid Nitrogen free Bromo thymol blue media. ASI shall show effective root development on all crops against which the inoculant is intended to be used. There should be 10 % increase in the plants over the dry mass on uninoculated control.
8.	Marking on Packet	Each packet shall be marked legibly to give the following: a. Name of the product b. Name and address of manufacture c. Name of carrier

		d. Batch number e. Date of mfg. f. Date of expiry g. Net mass h. Crops for which intended i. Storage instructions.

PHOSPHATE SOLUBLISING INOCULANTS-BACTERIAL (PSBI)
ISI STANDARDS (IS 14807-2000)

1.	Viable Cell Count	PSBI shall contain 10^7 viable bacterial cells/g of the carrier material or 10^7 cell/ml of liquid material before expiry.
2.	pH	6.5-7.5
3.	Moisture %age	35-40%
4.	Shelf Life	Not less than 6 months
5.	Carrier Material	Peat, lignite, charcoal, capable of passing through 150-212 micron IS sieve.
6.	Contamination	Nil at 10^5 dilution
7.	Efficiency character	PSBI shall have phosphate solubilising ability in the range of 30-50% and in terms of zone formation, minimum 10mm solubilization zone.
8.	Marking on Packet	Each packet shall be marked legibly to give the following: a. Name of the product b. Name and address of manufacture c. Name of carrier d. Batch number e. Date of mfg. f. Date of expiry g. Net mass h. Crops for which intended i. Storage instructions.

Note: For further information refer FCO Amendment order dated 24th March 2006 Schedule III (see clause 2(h) and (q) part A to D

—•—

Chapter 11

Field Application of Biofertilizers

Biofertilizers have to be properly applied to the seed or the soil. For this, appropriate application guidelines are available. First step is to purchase the right type of biofertiliser of good quality from a reputed source and use it before the expiry date It should be freshly prepared. Some hints while buying biofertilisers are:

- Different crops require different biofertilisers. Be certain to have the right one. The name of the crop should be mentioned on the packet. For pulses and other legumes, only *Rhizobium* of the right type for that crop should be used. Just any type cannot be used. It can not to be used for other crops.
- Be certain that the inoculant is fresh. Look for the expiry date after which the inoculant cannot be expected to produce good results, as it may not contain required number of bacteria. Use before expiry date.
- As far as possible, use products made by a reputed manufacturer with ISI mark.
- The package should provide clear instruction and the name and address of the manufacturer should be clearly printed on the bag.
- Never use any product, which has expired. For carrier based products, see that the product is having sufficient moisture. Product, which had dried, should not be used.

RHIZOBIUM

For Field Crops

- Dissolve 50 g country sugar (gur or shakkar) in half liter (500 ml.) water and heat for 15 minutes. Some producers directly add gum acacia as adhesive in the product. In such cases the addition of gur or shakkar is not needed. Farmers are advised to follow the instructions on the packet.
- Add 200 g gum Arabic to the solution and cool to room temperature.
- Mix biofertiliser (200 g) in it to form slurry.
- Add seeds required to plant one acre (0.4 ha) to inoculum slurry.
- Mix seeds with inoculum slurry by hand.
- Dry seeds in shade on a plastic sheet/paper.
- Seeds are ready for being sown. Sowing during hot period of the day should be avoided. The amount of culture, water and sticker needed/ha depends on the seed size and seed rate because the objective is to coat/cover the seed with the biofertiliser slurry. There should be a minimum 24-hour gap between seed treatment with a fungicide and biofertilizer treatment to avoid any harmful effect of the agrochemical on the biofertilizer.

An overall idea of the suitable quantities of inoculant (biofertilizer) and sticker for various legumes is provided.

Table 11.1. Suitable quantities of *Rhizobium* inoculant and sticker for inoculating legume seeds

Legume	Seed wt (g)	Inoculant (g)	Gum Arabic Solution (ml)
Groundnut	100 g	10 g	4.0
Chickpea	100 g	7 g	3.5
Pigeon pea	100 g	8 g	3.5
Soybean	100 g	10 g	3.0
Lentil	50 g	5 g	2.0
Subabul	50 g	10 g	3.0
Green gram	100 g	9 g	3.5
Cowpea	100 g	8 g	3.5

Tree Crops

Tree legume seedling can be readily inoculated in the nursery. A 50 gram bag of inoculant has the capacity to inoculate 10,000 seedlings (regardless of the species). Mixing the culture in cool water and using the suspension to irrigate the rooting medium of the seedlings can easily deliver this. The relation between number of seeds per kg and rhizobia per seed is given in Table.

Table 11.2. Relationship of seed number/kg to the number of rhizobia for inoculating some forest tree legume seeds and seedlings with rhizobia

Tree Species	Common Name	Number of Seeds per Kg.	No. of Rhizobia per Seed*	Nature of Rhizobia
Acaccia auriculiformis	Bengali babul	100,000	1.5×10^6	Slow growing
Gliricidia Sepium	Madre tree	8,100	18.5×10^6	Fast growing
Leucaena leucocephala	Subabul	14,000	10.7×10^6	Fast growing
Pithecellobium dulce	Vitali Imli	17,000	8.8×10^6	Fast growing

Based on 50 g. Inoculant/kg of seed and density of 3×10^9 rhizobial cells/g carrier.

While explaining the method of application, the following precautionary measures can be mentioned.

Do's	Don't
Keep biofertilizer packets under shade.	Don't use poor quality biofertilizer.
Mix biofertilizer with seeds as per full requirement.	Don't use packets on which crop, date, batch no., name of manufacturer and expiry date is not mentioned.
Dry treated seeds in shade.	Don't mix biofertilizer with insecticides, fungicides, herbicides or fertilizer.
Sow treated seeds immediately after drying. The better time of sowing is either the early morning or afternoon when heat is tolerable.	Don't mix the biofertilizer with seed under direct sunlight.

AZOTOBACTER

- Make a slurry of the carrier based biofertilizer using minimum amount of water.
- Dry the seeds in shade and sow.
- For transplanted crops, dip the roots of seedlings in the above slurry for 20-30 minutes and then transplant the seedlings.
- In case of sugarcane, *Azotobacter* application may be needed more than once during early growth. In that case, second and further treatments can be given by pouring the slurry near the root zone. The biofertilizer can also be mixed with farmyard manures and broadcasted close to the root zone.

AZOSPIRILLUM

Application technique is similar to that described for *Rhizobium*. For transplanted rice, the roots of rice seedlings can be dipped in the biofertilizer slurry for 20-30 minutes before transplanting.. For pearl millet and finger millet, an application rate of 2 kg/ha of *Azospirillum* is recommended.

For rice nursery also 600 g/ha *Azospirillum* culture is recommended. It is to be mixed with water in which seeds are soaked overnight before sowing in the nursery. In the main field, a further application of 2 kg *Azospirillum* (10 packets)/ha is recommended. It is to be mixed with 25 kg FYM + 25 kg soil and broadcast over the field before transplanting. Another method recommended is to prepare a slurry of 1 kg *Azospirillum* biofertilizer in 40 lit water and dip the roots of rice seedlings in it for 15-30 minutes before transplanting.

PHOSPHOBACTERIA

For the application of Phosphate Solubilisers, the best method is seed treatment. Other methods like seedling treatment and soil treatment are also in practice.

Seed Treatment

The method of inoculation of seed with microbial inoculnat affects their efficacy. In general, the inoculant is mixed in clean water or in 5% solution of an adhesive to make thick slurry. Many a times biofertilisers manufacturers add gum Arabic in their product, in this case there is no need to use any adhesive. This slurry is then mixed with the required quantity of seeds to be sawn in an acre of land. The seeds are dried in shade and are stored in cool dry place before use. Generally about 25- 30g inoculant is required to coat 1 kg seeds. However, exact quantity will depend upon size of the seed. Generally large seeded crop viz; rajma, groundnut etc. require 1.5 kg inoculnat /ha, while medium sized seed viz: gram, soybean and small sized seeds viz; moong bean, urid bean cowpea, Lucerne etc. will need 1.0 or 0.5 kg inoculnat, respectively. Seeds should be sown as soon as possible after treating with inoculants to take full benefit of the inoculation.

The seeds can also be inoculated by sprinkle method. In this case the adhesive solution is first mixed uniformly with the seeds and extra solution, if any should be decanted from the seeds. Contents of inoculant packet are then sprinkled over the seeds and mixed uniformly with the seeds.

The methods treating seed with inoculants do not hold good, if the agro climatic conditions are not favorable. To protect the inoculated seed from soil stresses, it is necessary to coat the inoculated seeds with inert materials. This technique is known as pelleting. Pelleting provides prolonged survival to the rhizobia on seeds. The most commonly used coating materials are calcium carbonate for acidic soils and gypsum for saline neutral soil conditions.

- Prepare the slurry of 200 g of biofertilizer in 200-500 ml of water.
- Pour this slurry slowly on 1-25 kg seeds. Mix the seeds evenly with hands to get uniform coating on all seeds.
- Dry the treated seeds in shade and sow immediately.

Method of seed pelleting

- *Preparation of the adhesive and inoculant mixture:* Take preservative free, light coloured and non-toxic gum Arabic for making adhesive solution. Dissolve 100 g gum Arabic in 225 ml hot water with mechanical mixing. Neutralize the solution with required quantity of calcium carbonate. Thoroughly mix 100 g of carrier-based inoculum with the solution, which upon swelling of solids becomes 350 ml approximately. This quantity of slurry is sufficient to inoculate 7 kg of small, 14 kg of medium or 28 kg of large sized seed.
- *Forming pellet:* Mix the adhesive and seed in a mechanical shaker until the seeds are evenly coated. The above mentioned amounts

of inoculated seed require 3.5 kg of calcium carbonate as pelleting materials, which should then be added in and mixed rapidly until all the seeds are evenly coated (usually 1-2 minutes are required).

If pellets are powdery or soft, either too much lime has been added or the mixing has been uneven. Pasty-looking pellets with some seed surface exposed indicate too much adhesive solution; adding more lime may rectify this. Clumping of small seeds may be caused by the addition of too much adhesive and these may not break down on adding the lime to become very compact and to crack and flake off when dry. At present many biofertilizer manufacturer add 1-2 : gum arabic in final product , so there is no need to use any extra sticking agent.

Seedling Treatment

- Prepare a suspension of 1-2 kg of biofertilizer in 10-15 liters of water.
- Dip the seedlings from 10-15 kg of seeds into the suspension for 20-30 minutes and transplant treated seedlings immediately.

Soil Treatment

- Prepare the mixture of 2-3 kg of biofertilizer in 40-50 kg of soil/ compost.
- Broadcast the mixture in one acre (0.4 ha) either at sowing or 24 hours earlier.
- The broadcast may be done in standing crops just before irrigation
- 4-5 kg Biofertilizer in 120-150 kg of soil/compost should be used for long duration crops.
- Do not mix chemical fertilizers, insecticides or pesticides directly with Biofertilizers.

— • —

Chapter 12

Liquid Biofertilizers: A New Vistas

Liquid biofertlizers are the microbial preparations containing specific beneficial microorganisms, which are capable of fixing or solubilising or mobilizing plant nutrients by their biological activity. They are broadly classified into two groups.

- ☆ Nitrogen fixing microorganisms
- ☆ Phosphorous solubilising/mobilizing microorganisms

Among the two groups Phosphorous solubilising microorganisms and *Azotobacter* inoculant are widely used compared to other biofertlizers.

LIMITATIONS OF CARRER BASED INOCUNANTS

1. Shorter shelf life (Usually 6 moths).
2. Lack of identifiable character Sensitivity to temperature.
3. More chances of contamination.
4. Sensitivity to temperature.
5. Poor moisture retention capacity.
6. Lack of instant visual effects on application.
7. Restriction on use as a measure of conservation (wood charcoal).
8. Unavailability of good carrier in local area.
9. Quality inconsistency of carriers.

10. Bulk sterilization problem in term of economy and facilities.
11. Labour intensive.
12. Poor cell production.
13. Even less contamination in packets disturbs enzymatic activity of beneficial bacteria.
14. Problem of proper packing.
15. High transport cost.
16. Takes more time for quality control.
17. Dosages contradictory to farmers.
18. Very slow adaptation by farmers.
19. Less commercial value and less scope for export.

LIQUID FORMULATIONS

Although, India boasts to be the largest biofertilizers producer in the world, the creditability of biofertilizers is slow among farmers. Due to limitations of the carrier-based inoculums, formulation of liquid based biofertilizer was done. This technology is an alternative solution to carrier based biofertilizers, microbial pesticides and bio-control agents and this is not the usual broth culture from the fermentor or a water suspension of the carrier based biofertilizers.

What is liquid biofertilizer?

Liquid biofertilizers are special liquid formulations containing not only the desired microorganisms and their nutrients, but also special cell protectants or substances that encourage formation of resting spores and cyst for longer shelf life and tolerance to adverse condition.

BASIC CONCEPTS OF LIQUID FORMULATION

There are 4 basic characteristics in formulation.

They are:

1. To stabilize the organisms during production, distribution and storage.
2. To easily deliver in the field in the most appropriate manner.
3. To protect the microorganisms from harmful environmental factors at the target site (field), thereby increasing persistence.

4. To enhance activity of the organism at the target site by increasing its activity, reproduction, contact and interaction with the target crops.

ADVANTAGES OF LIQUID BIOFERTILIZERS

The advantages of liquid biofertilizers over conventional carrier based biofertilizers are listed below:

Number of advantages noted by the authors from last 6 years, and some are mentioned here:

1. Longer shelf life 12-24 months and high population can be maintained more than 10^8 cells/ml upto 12-24 months.
2. No contamination.
3. Less effect of high temperature
4. No loss of properties due to storage upto 45°C.
5. Greater potentials to fight with native populations.
6. Identified by typical fermented smell.
7. Cost saving on carrier material, pulvcrization, neutralization, sterilization, packing and transport.
8. Quality control protocols are easy and quick.
9. Better survival on seed and soil.
10. No need of biofertilizer production units through out the year.
11. Very much easy to use by the farmer.
12. Dosages are 10 times less than carrier biofertilizers.
13. High commercial revenues.
14. High export potential.
15. Very high enzymatic activities since contamination is nil.
16. Can be used in drip irrigation in foliar spray and green house.

Further, it is very necessary to understand that, liquid biofertilizers are not merely usual broth cultures from fermentors, which are specially prepared with desired agriculturally important microorganisms along with proper nutritional base to facilitate protection of organisms from adverse conditions and resting spore formation or cysts etc.

In India, Liquid formulations of biofertlizers are marketed by

1. Kumar Krishi mitra, Pune
2. Hindustan Antibiotics, Pimpri, Pune.
3. Agriland Biotech, Samlaya, Vadodara

Recently, Anand Agricultural University has also developed protocol for the liquid formulations of Phosphate culture and *Azotobacter.*

—•—

Chapter 13

Project Profile for Setting of Biofertilizer Production Unit (150 MT/Year)

INTRODUCTION

Biofertilizer are carrier based microbial inoculants containing cells of efficient strains of microorganisms. They are used for enhancing the productivity of the soil. They fix atmospheric nitrogen and solublize phosphorus. Also they stimulate plant growth through synthesis of plant growth promoting substances.

The low cost, high efficiency technology of Biofertilizer application has an important role to play in increasing agricultural production. With rapid depletion of non-renewable energy. Role of Biofertilizer in the agriculture enterprise becomes more important as it economizes the fertilizer application to corps. Application of Biofertilizer has thus, become integral components of Integrated Nutrient Management System.

PRODUCTION CAPACITY

The proposed unit will produce 150 m.t. of bio-fertilizer at the rate of 19 m.t. per month and will also produce microbial pesticides.

The unit will produce the following bio-fertilizer:

1. *Azotobacter* for all cereal crops.
2. Phosphate culture for all crops.
3. *Azospirillum* for cereal crops
4. *Rhizobium* for pulses.
5. *Acetobacter* for sugarcane.

MARKET POTENTIAL

There is huge demand for Biofertilizers in India and Gujarat in particular. (5,000 M.T. approximately), provided proper awareness is created among the farmers. The awareness is increasing among the farmers for organic farming now.

RAW MATERIALS

1. Source of Nutrients
2. Minerals & salts
3. Plastic bags / Carton Boxes
4. Lignite
5. Calcium carbonate
6. Gum arabic

13.1. Mass multiplication of Biofertilizers Basic assumption

1	Total land necessary	One ha. Approx.
2	Construction area	955 sq.m.
3	Total investment – Phase I	279 lacs
4	Production capacity	150 t/annum
5	Requirement of raw material carrier	90 ton
6	**Manpower Requirement**	
	Project manager	1
	Senior Microbiologist	1
	Assistant Microbiologist	2
	Senior Research Assistant	2
	Junior Research Assistant	3
	Agricultural Assistant cum field man	2
	Laborers	5
Note: Manpower would be engaged on contractual services.		
7	**Estimated Average Prices**	**Rs./pkt**
	Azotobacter	12
	Azospirillum	12
	Rhizobium	12
	Phosphate culture	12

Contd...

Table 13.1. Contd...

8	**Estimated Production/year**	**T/year**
	Azotobacter	37.5
	Azospirillum	37.5
	Rhizobium	37.5
	Phosphate culture	37.5

Table 13.2. Name of strain: *Azotobacter chroococcum*

S. No	**Details of production**	
1.	Total Production capacity (1 packet = 200 g)	150 t/year (75,0000 packets)
2.	No. of working days	240 days (8 months)
3.	Average daily production capacity	3124 packets (624 kg)
4	No. of shift	1 (8 hours)
5	No. of daily wages labour required	6.0 (@ 500 packets/labour)
6	Requirement of mother culture	1 %
7	Requirement of broth per day	220 lit
8	Requirement of carrier per day	416 kg
9	Broth required for one tonne	400 liter
10	Carrier required for one tonne	600 kg
11	Sievability of carrier	100 mesh (150 μ)
12	Mixing ratio (carrier: broth)	6.0: 4.0 (W: V)
13	Time required for sterilization	
	Carrier	4 hours
	Broth	45-60 minutes
14	Carrier required for 1000 packets	120 kg

13.3. Requirement of chemicals for production of fermented broth / 1000 pkts.

A	***Azotobacter* broth/1000 pkts**	
	Chemical	**Gram**
	Mannitol	1400
	Dipotassium hydrogen phosphate	14

Contd...

Table 13.3. Contd...

	Magnesium Sulphate	14
	Sodium Chloride	14
	Potassium Sulphate	7
	Calcium Carbonate	350
B	***Azospirillum* broth/1000 pkts**	
	Maleic acid	350
	Potassium hydroxide	280
	Dipotassium hydrogen phosphate	35
	Ferric Sulphate	3.5
	Manganese Sulphate	0.7
	Magnesium Sulphate	7.0
	Sodium chloride	1.4
	Calcium chloride	0.7
	Sodium molyblate	0.14
C	***Phosphate* culture broth/1000 pkts**	
	Glucose	700
	Tri-calcium diphosphate	350
	Diammonium Sulphate	35
	Potassium chloride	14
	Manganese Sulphate	7
	Magnesium Sulphate	1
	Ferric Sulphate	1
	Yeast extract	35
D	***Rhizobium broth*/1000 pkts**	
	Dipotassium Phosphate	35
	Magnesium Sulphate	14
	Sodium Chloride	7
	Mannitol	700
	Yeast Extract	70

Table 13.4. Requirements: Building Space requirement for biofertilizer production unit Annual Production Capacity: 150 t/annum

S. No.	Name of unit	Size (Sq. m.)	Facilities required
(A) Inoculum wing			
1.	Storage room	50	Racks, Cabinets etc.
2.	Media preparation room	50	Vertical autoclave, pH Meter, Distillation plant, Deioniser, Oven, sink, Lab. Bench etc.
3	Bacteriological room	50	Incubator, Fridge
4.	Fermentation room	75	Fermentors, /Rotary shakers, Lab bench, Air conditioner etc.
5.	Quality control room	45	Lab bench, sink, microscope, pH meter, Colony counter, Spectrophotometer, Balance
6.	Growth room	30	Slotted angle rack with illumination, Leonard jars, air conditioner, timer etc.
(B) Carrier wing			
7.	Store room for carrier	50	-
8.	Carrier sterilization room	40	Horizontal Autoclave, trolley, Sink, Lab Bench, Trays etc.
9.	Mixing room	45	Tray, Bench
10.	Curing room	35	Air conditioners Temp. & Humidity controller, sealing machine, corrugated boxes etc.
11.	Packing room	30	
(C) Customers wing			
12	Office room	100	-
13.	Biofertilizer Storage room	50	-
14	Selling room	15	Air conditioners
15	Net House	130	Bench, sink earthen pots etc.
	Total	795	

Total expenditure: Rs. 23.85 lacs (795 sq. m @ Rs. 3000/sqm construction cost)

13.5. Requirement (Rs. Lacs): Equipments

Equipments	Cost/item	Units Reqd.	Total Amt. Reqd.	Justification
Research Binocular Microscope with CCTV	2.00	1	2.00	Study of microorganisms
Two tier Rotary Shaker (200-250 lit)	1.70	2	3.40	Mass production of biofertilizers
Horizontal Autoclave 0.9 x 0.9 x 1.8 m	5.00	1	5.00	Sterilization of media, glassware, carrier etc.
Refrigerator 300 lit	0.30	2	0.60	Storage of microorganisms, chemicals etc.
Computer with Accessories	0.50	2	1.00	Data compilation and analysis
Deep Fridge	0.30	1	0.30	Plant samples
Portable Shaker	1.35	2	2.70	Preparation of starter culture
Laminar Airflow 180 x 60 x 60 cm	1.00	2	2.00	For providing the microorganisms free atmosphere
Autoclave vertical	0.25	2	0.50	Post sterilization of media
Electronic balance	0.25	2	0.50	For weighing chemicals etc.
Mixing machine	2.00	1	2.00	For mixing of biofertilizer product
Distillation Plant -Quartz	0.30	1	0.30	For getting pyrogen free distilled water
Air curtain	0.20	2	0.40	For providing microorganisms free air.
Colony Counter-	0.20	1	0.20	For quality control
pH meter	0.10	2	0.20	For adjustment of pH
Bag sealing machine Manual	0.20	2	0.40	For sealing of plastic bags.
Hot air oven	0.30	1	0.30	Drying and sterilization of glassware
Colorimeter	0.40	1	0.40	For quality control

Contd...

Table 13.5. Contd...

Air conditioner 1.5 ton	0.40	3	1.20	For providing controlled conditions in laboratories
Fermenter 100 - 200 lit	4.00	2	8.00	Large-scale fermentation of biofertilizer strains.
Green House	3.00	1	3.00	For conducting pot culture study -Quality testing of products
Portable Genset	3.00	1	3.00	For providing electricity during power failure
Other small equipments costing less than Rs. 10000/-	0.50	-	0.50	-
Total			37.90	

Rupees thirty-seven lacs and ninety thousand only.

13.6. Requirement: Manpower

S. No.	Details Manpower on contractual services	Fixed Pay @ Rate Rs./PM	Year				
			Post	I	II	III	Total
			Rs (Lacs)				
1.	Senior Microbiologist	10000	1	1.20	1.20	1.20	3.60
2.	Assistant Microbiologist	7000	2	1.68	1.68	1.68	5.04
3.	Senior Research Assistant	5000	2	1.20	1.20	1.20	3.60
4.	Junior Research Assistant	4000	3	1.44	1.44	1.44	4.32
5.	Agriculture Assistant cum Field man	4000	2	0.96	0.96	0.96	2.88
6.	Laborers	2500	5	1.50	1.50	1.50	4.50
	Grand Total		15	9.30	9.30	9.30	23.94 23.90

Rupees twenty-three lacs and ninety thousand only.

13.7. Cost/Benefits Aspect of the Product

S. No.	Item Required (For 150 tonnes/annum)	Rate (Rs.)	Amount (Rs.) Lacs
A	**Fixed cost**		
	Building = 955 m^2 @ 5000/m2		23.85
	Equipments		37.90
	Glass wares & miscellaneous items		3.00
	Total		64.75
B	**Variable cost**		
	(a) Row materials cost		
	Carrier (90 tonnes)	4500/tonne	4.05
	Broth (60,000 lit)	10.00/lit	6.00
	Gum Accacia (1500 kg.)	120/kg	1.80
	Polythene bags (7,50,000)	0.70/pkt.	5.25
	Carton box (50 pkts.) 15, 000	15.00/box	2.25
	(b) Publicity cost	0.30/pkt	2.25
	Total		21.60
C.	**Indirect cost**		
	Salary and wages		9.3
	Diesel, Maintenance of vehicle		3.0
	Office expenses including transport, stationary etc.		3.0
	Total	-	15.30
	Thus for 7,50,000 packets, the production cost will be		
	Variable cost		21.60
	Indirect cost		15.30
	Depreciation on fixed asset @ 10% fixed cost		6.47
	Power consumption (0.080 KW/pkt @ Rs. 5.00/unit		3.00
	GRAND TOTAL		46.37
	Total cost of production of 750, 000 packets		46.37

Contd...

Table 13.7. Contd...

	Therefore, the cost of one packet		6.18
	Production cost / pkt		6.20
	SALE PRICE PER PACKET		
	Production cost per packet		6.20
	Risk coverage for unsold packets		0.60
	Coverage for freight & transport cost		1.00
	Adjustment on variability on input cost		0.50
	Approx. cost / packet		8.30
	Profit margin / pkt		2.20
	Commission to wholesaler / retailer		0.50
	RETAIL SALE PRICE / PACKET		11.00

13.8. Total Financial Outlay (Rs. Lacs)

Details of expenditure	**Year**			
	I	**II**	**III**	**Total**
(A) Recurring				
1. Salary and manpower	7.97	7.97	7.97	23.90
2. Chemicals	2.00	2.00	2.00	6.00
3 Lignite carrier	1.01	2.02	4.05	7.08
4 Gum Arabic	0.45	0.90	1.80	3.15
5 Broth	1.50	3.00	6.00	10.5
6 Plastic wares	1.31	2.62	5.25	9.18
7 Polythene packets	0.56	1.12	2.25	3.93
8 Corrugated box	0.56	1.12	2.25	3.93
9 Glassware	2.50	2.50	2.50	7.50
10 Cost on hiring on vehicles	2.00	2.00	2.00	6.00
11 Publicity	0.56	1.12	2.25	3.93
12 Office expenditure	2.00	2.00	2.00	6.00
13 Travel and training	2.00	2.00	2.00	6.00

Contd...

Table 13.8. Contd...

(B) Non recurring				
Equipments	37.90	-	-	37.90
Buildings	23.85	-	-	23.85
Furniture	3.00	2.00	-	5.00
(C) Working capital	20.00	-	-	20.0
Total A+ B+ C				183.85
			i.e.	184.00

Rupees One Eighty Four lacs only

I year Production = 37.5 tonnes /annum

II year Production = 75.0 tonnes /annum

III year Production = 150.0 tonnes /annum

— • —

Chapter 14

Microbial Pesticides An Alternative Tool for Management of Agriculture Pests

The advent of insecticides with unrestrained applications of wide range of pesticides created several complications in the agro ecosystem, which has forced the agriculturists to adopt alternative management practices, in which use of microbial pesticides is well versed. Pest problem is one of the major constraints for achieving higher production in agriculture crops. India loses about 30% of its crops due to pests and diseases each year. The damage due to these is estimated to be Rs.60, 000 crores annually. The use of pesticides in crop protection has certainly contributed for minimizing yield losses.

During recent years the concept of integrated pest management has arised arise but still farmers have continued to rely heavily on chemical insecticides for pest control. The intensive and indiscriminate use of insecticides for crop protection has created problems such as toxic residue in and on the crop plants, agro-ecosystem pollution etc. Though the average quantum of pesticides usage in India is low, the damage caused due to their indiscriminate usage and poor quality maintenance is alarming. Interims of value, much of the pesticide application is accounted for few crops. For example, cotton, paddy and vegetable crops account for 80% of the value of pesticides applied in India. Pesticides or chemicals are meant to control harmful pests such as insects, nematodes, diseases, weeds etc. However, excessive use of pesticides not only leave residues in soil, water and air but also have adverse effects on the non target organisms such as pollinators, parasitoids, predators and wild animals. The average consumption of pesticides in India is about 570 g per ha. compared to developed countries like Japan and Germany where the consumption rate is 11 kg and 3 kg per ha, respectively. This has adversely affected the ecological balance resulting in pest resurgence and development of resistance in the pest species which has resulted in high cost of production and low income especially to cotton farmers in Andhra Pradesh and Maharashtra. Wide application of chemical pesticides on agricultural crops is generally harmful to beneficial insects and animals, which feed on and are also destructing natural fauna and flora. In order to overcome these problems, biological control is one of the safe approaches for pest management. Insect control through an eco-friendly manner is no longer a dream because many biological, botanical and natural pesticides are promoted successfully in the world market by now. Insect pathogens are effectively demonstrated for pest control on different crops, since the Second World War.

The chemical insecticides are valuable for control of insects, but due to their continuous and over usage during last few decades have posed several serious problems.

- Development of resistance in pests.
- Resurgence of minor pests.
- Toxicity in the environment, accumulation of pesticides residues.
- Biological magnification of pesticide residues through the food chain.
- General environmental pollution

In view of the several disadvantages associated with the unscientific use of pesticides in agriculture, there is an urgent need for minimizing the use of chemical pesticides in the management of insect pests. Growing public concern over potential health hazards of synthetic pesticides and also steep increase in cost of cultivation or low profit making by farmers has led to the exploration of eco-friendly pest management tactics such as Integrated Pest Management (IPM). IPM aims at suppressing the pest species by combining more than one method of pest control in a harmonious way with least emphasis on the use of insecticides. In simple terms "IPM is the right combination of cultural, biological and chemical measures which provides the most effective, environmentally sound and socially acceptable methods of managing diseases, pests and weeds". The major components of IPM are prevention, observation and intervention. The IPM seems to be the only answer to counter some of the major pests of crops, which have become unmanageable in recent years. The success of IPM largely depends upon conservation of naturally occurring bio control agents and timely application of bioagents in field.

Bio-pesticides

In nature every ecosystem exists in a balance. Growth and multiplication of each organism depends on the food chain, its predators, parasites, etc. In biological control system, these interrelations are exploited. The natural enemy of a pest, disease or weed is selected, its biology is studied for mass multiplication and utilize the same to check the target pest. They are also specific in their action and perish once their feed (i.e. the pest) is exhausted. Thus they are based on natural principles; do not leave any residue, safe and economical. Among the alternatives, biological control of pests is one of the important means for checking pest problems in almost all agro-ecological situations.

Biopesticides are living organisms, which can intervene the life cycle of insect pests in such a way that the crop damage is minimized. The agents employed as biopesticides, include parasites, predators and **disease causing fungi, bacteria and viruses**, which are the natural enemies of pests. Further, they complement and supplement other methods of pest control. Utilization of naturally occurring parasites, predators and pathogens for pest control is a classical biological control. On the other hand, these bio agents can be conserved, preserved and multiplied under laboratory condition for field release. Once these bio-agents are introduced in the field to build their population considerably, they are capable of bringing down the targeted pest' population below economic threshold level (ETL). However, the crux lies in their mass production and application at the appropriate time.

ADVANTAGES OF BIOPESTICIDES

Bio-pesticides are preferred over chemical pesticides for the following reasons:

- No harmful residues.

- Target specific and safe to beneficial organisms like pollinators, predators, parasites etc.
- Growth of natural enemies of pests is not affected, thus reducing the pesticide application.
- Environmental friendly.
- Cost effective.
- Important component of IPM as 1st line and 2nd line of defense, chemicals being the last resort.

Last decade has witnessed a tremendous breakthrough in this aspect, especially on standardization of production techniques of ***Trichoderma, Gliocladium, Paecilomyces, Pseudomonas, NPV, Beauveria, Metarhizium, EPNs and Bacillus*** spp. to use them against many insect pests and diseases. The popularity of biopesticides has increased in recent years, as extensive and systematic research has greatly enhanced their effectiveness. Also, techniques for the mass production, storage, transport and application of biopesticides have been improved in recent years.

Commercial Production of Biopesticides

Though there are many biopesticide production units existing in the country, as on today, they are able to meet the demand of only less than 1% of cropped area. There exists a wide gap, which can only be bridged by setting up of more and more units for production of biopesticides. This requires large-scale investment and private participation. There is a scope to enhance production and use of biological control agents in the days to come as the demand is on the increase every year.

FEW MICROBIAL PESTICIDES MARKETED IN INDIA

More than 25 products are now available in India out of which majorities are produced by small-scale regional producers and few products are also from State Agricultural Universities, but majority of them are not registered products OR having provisional registrations (Table 14.1)

Characteristics for development of A MICROBIAL BIOAGENT:

- *Efficacy:* A highly effective biocontrol strain or other material must be obtained or produced. Such strains must not only have

Table 14.1. *Microbial Pesticides* Products Available in local Market

S. No.	Name of Agent/s	Common Product Examples
1	*Bacillus thuringiensis*	Delfin, Halt, Dipel, Biolap etc.
2	Nuclear polyhedrosis virus (NPV)	*Ha*NPV, *Sl*NPV
3	Muscardine fungi, *Beauveria bassiana* and *Metarhizium anisopliae*	Basina, Biosoft, *Metarhizium, Beauveria* spores
4	*Trichoderma viride* *T. harzianum*	Tricho-XP, Mnitor WP, Monitor S, Trichomycil, T-2 Trichoderma spore powder and granules, etc.
5	*Pseudomonas fluorescence*	*P. fluorescence* wattable powder
6	*Paecilomyces lilacinus*	Yorker, Tcicho X-P (Combi product)
7	Entomopathogenic nematodes (EPNs)	Green commando, Soil commando, *Steinernema* sp.

appropriate mechanisms for biocontrol such as those just described, but it should also (a) be able to compete and persist in the environment in which it must operate, and (b) ideally, be able to colonize and proliferate on existing and newly formed plant parts at times well after application. But better to use existing natural strains, which should be sufficiently effective.

- *Cost:* Inexpensive production and formulation of the biocontrol agent or other material in question must be developed. The production process must result in biomass with excellent shelf life even under adverse storage conditions. In many respects, the requirements for production of products for agricultural use are more difficult to meet than those required for pharmaceutical products. If agricultural materials are to be successful, they must be very inexpensive, able to be produced in large quantities, and maintain good viability without specialized storage systems. In my opinion, the largest technological impediment to the use of biocontrol agents is the lack of both technology and facilities for biologically- and economically-effective production of biocontrol organisms. Nearly no public-sector research or funding is directed towards research on methods of production of biocontrol organisms.

- *Application:* Delivery and application methods that permit the full expression of the biocontrol agent. Delivery systems must ensure that biocontrol agents will grow well and achieve their purpose. In our experience, delivery and application processes must be developed on a crop by crop and application by application basis. No general solutions exist, and so biocontrol systems must be developed for each crop. An analogy may be helpful to understand these requirements. In many cases, biocontrol researchers have attempted to use our biocontrol agents in a manner analogous to the sowing of crops. An effective method of use is to use the biocontrol agent in conjunction with chemical pesticides. The chemicals provide good short-term protection, and the biocontrol provides long-term protection. As a consequence, yields frequently are increased over use of the chemical alone.

  It can also be used as an additive to greenhouse potting mixes. It can both reduce pesticide use in the greenhouse by limiting root-attacking diseases and protect transplants in the field by virtue of its ability to colonize roots.

  When added to turf, the granular formulation results in establishment of the organism on roots. However, the pathogens may survive in sufficient numbers to cause disease. Once the pathogens are established on the foliage, the soil-applied bio control agent no longer can protect the plant. This granular formulation can therefore result in disease reduction, but it must be used in conjunction with compatible chemical fungicides. A spray application of the bio control agent for control of foliar phases of the diseases is also developed recently and has provided excellent results in field tests.

  The organism is also highly effective when applied to blossoms or fruits for control of diseases. Even low levels of the organism applied to strawberry blossoms by bee delivery or by sprays of liquid formulations are effective. For maximum control this initial application needs to be augmented by sprays as fruits mature, and addition of a tank mix to this late application appears to have synergistic activity over either the bio control agent or the chemical pesticide alone.

- *Commercialization:* Once an effective bio control system has been identified, it can and should be commercialized as discussed above. As a bio control researcher, it has become clear that solving the technological problems is perhaps the easiest part of developing bio control agents and systems. Legal and commercial considerations are also extremely important. Over the past 15 years, a series of filed trials through the State Agricultural Universities that covered the relevant technology in India. Protection of intellectual property was essential to further development. If the technology cannot be protected, few companies may be interested in developing it further. At this time, a large national company is interested in adding an agricultural biotechnology line to its market mix and acquired the technology from University. They immediately began efforts to develop large-scale production systems using their liquid fermentation facilities and to register the product. This company acquired rights to the technology from my lab, and is designed to develop, produce, and market bio control products for agriculture Thus Bio control system, once they are developed can be commercialized, but only if adequate business, legal, and financial expertise, and large-scale production techniques and facilities are available or can be constructed.

BASIC REQUIREMENTS FOR ESTABLISHMENT OF BIOPESTICIDE UNITS

Based on the functioning bio-control production units and in line with the technology and objective of bio pesticides production, various facilities required for the successful implementation of such projects are indicated below:

Land

Land is required for construction of culture and rearing rooms, processing room, laboratory, office etc. but if only rented buildings, no land cost has been considered except for poly house.

Building and Civil Works

Biopesticides production involves an estimated built up area of about 1000 sq to 2400 sq. ft. area is required. Other utilities required are power, water and vehicle.

Plant and Machinery

There is no requirement of heavy plant and machinery. Apart from this centrifuge, mixers, shakers, autoclave and some fabricated equipments, fermentors, laminar flow apparatus etc. for microbial agents are required. All the machinery required are locally manufactured.

Raw Material

For large scale production of fungal bioagents a molasses-yeast medium, is required. All these materials are available locally.

Water

The water requirement is mainly for feed preparation, washing, cleaning, drinking etc. Water quality should be tested to establish the suitability.

Power

Power supply is essential for bio-pesticide units. Electricity charges under recurring cost are considered in the models.

Manpower

Production of bio-pesticides required skilled manpower. Technical staff skilled labors and semi-skilled labors.

Scale of Production

These biopesticides can be produced on a small or large scale. Small-scale production is particularly suitable to village or community level cooperatives, which can produce and distribute these for local use. As the production technology of some of these agents (particularly *Trichoderma*) is relatively simple, the local farmers or SHGs can be trained to undertake the production. Medium & large-scale production can be undertaken by firms, sugar mills cooperatives engaged in the manufacture and distribution of agro-chemicals. For example, fertilizer companies, which already possess sufficient in-house technological expertise and marketing resources, are ideally suited for producing bio pesticides on a large scale. Similarly, seed companies are particularly well placed for undertaking the production and marketing.

Market Potential

Considering the negative effects of indiscriminate case of pesticides, importance for organic farming and promotion of sustainable farming practices it is estimated that there will be further scope for new units, particularly in the states of Maharashtra, Gujarat, Rajasthan, Madya Pradesh, Tamil Nadu, Andhra Pradesh, UP, West Bengal and Karnataka, where crops such as sugarcane, cotton, pulses, cereals and vegetable crops are grown in large scale. Estimated the gross demand for a few biopesticides which is given below: At present, in some states, state government is purchasing the product from the private parties and selling it to the individual farmers at a subsidized rate.

BIOPESTICIDES REGISTRATION

At present, *Bacillus thuringiensis*, neem based formulations, microbial pesticides like fungi, NPV etc., are included in the schedule of Insecticides Act, 1968. This ensures the quality of bio-pesticides at farmers level. The standard parameters, protocols for data generation, guidelines for registration are prepared and circulated to prospective entrepreneurs by Govt. Now as such, any person dealing with bio pesticides without registration is ill legal. Accordingly, registration procedure is included in **Annexure V**.

FIELD EFFICACY OF BIOPESTICIDES

State Agricultural Universities and ICAR Research Institutes and Stations to know the extent of pest control by application of some common microbial biopesticides have conducted field efficacy trials during last few decades.

The percentage of pest control achieved for example:

Efficacy and pest control by two common biopesticides NPV up to 70-80%; *Trichoderma viride* up to 60-90% in different crops, reported from various states in India. More details are in chapter 15 to 20.

FUTURE THRUSTS IN INDIA

Due to chemical pesticide problems in India, there is an urgent need to promote environmental friendly bio- pesticides in the country. Moreover recent Government policies also favor Bio pesticide.

- Industrial fermentation technique for bioagents needs to be developed to launch biopesticides in the market.
- Large-scale field demonstrations at farmer's fields are required to increase awareness and adoption of bioagent for pest management.
- Use of effective bioagents should be accelerated in integrated management and economics needs to be worked out.

- Interest to be focused on developing new methods of biocontrol. In particular, identification of proteins, and the genes encoding them, from various microbial agents and newer microbes. These proteins degrade chitin, which is a structural component of pathogenic microorganisms and herbivorous insects. Investigation of large-scale production of the toxins and other gene products for the control of crop pests and diseases. This work will progress well in cooperation with researchers elsewhere in the world and should represent a next generation of bio control products.

—•—

Chapter 15

Trichoderma A Fungal Bioagent to Promote Growth and Fight Plant Diseases

Losses due to plant diseases are 10-20% of the total world food production every year, resulting in economic losses amounting to billions of dollars. Chemical control involves the use of chemical fungicides to reduce the pathogens or to protect the plants from infection by pathogens are very effective, but it is often non-specific in its effects, killing beneficial organisms and it may have undesirable health and environmental pollution risks. Biological control of diseases involves the use of living organisms to control the diseases that attack or interfere with the pathogens. The members of the genus *Trichoderma* are very promising against soil-born plant parasitic fungi. These filamentous fungi are very widespread in nature, with high population densities in soils and plant litters. They are saprophytic, quickly growing and easy to culture and they can produce large amounts of conidia with long lifetime. *Trichoderma* spp. are the fungi that are present in substantial numbers in nearly all agricultural soils and in other environments such as decaying wood. Among their other activities, they grow tropically toward hyphae of other fungi, coil about them in a lectin-mediated reaction, and degrade cell walls of the target fungi. This process (mycoparasitism) limits growth and activity of plant pathogenic fungi. In addition to, or sometimes in conjunction with mycoparasitism, individual strains may produce antibiotics. However, numbers and the physiological attributes of wild strains are not sufficient for highly effective control of plant diseases.

The antifungal abilities of these beneficial microbes have been known since the 1930s, and there have been extensive efforts to use them for plant disease control since then. However, they are only now beginning to be used commercially. Moreover, it is effective against a wide range of plant pathogenic fungi including *Pythium* spp., *Rhizoctonia solani, Fusarium* spp., *Botrytis cinerea, Sclerotium rolfsii* and *Sclerotinia homoeocarpa*. Crop losses due to soil borne plant pathogens worldwide are *Pythium* spp., *Furarium oxysporum, Sclerotium rolfsii, Rhizoctonia solani* and *Phytophthora* spp. These fungal pathogens generally cause wilt disease in many crops. *Trichoderma* spp. is commonly found in almost any soil and other natural habitats consisting of organic matter such as decaying bark, plant material, etc. They grow trophically towards hyphae of other pathogenic fungi, coil them and degrade their cell walls. In addition, they produce toxic metabolites, which protect the seeds from soil borne pathogenic fungi, by forming a protective coating on them. *Trichoderma* spp. is saprophytic fungi that grow best in neutral and acid soils and thrive well in moist conditions.

CHARACTERISTICS

Cultures are typically fast growing at 25-30°C, typically not growing at 35°C. Colonies at first transparent on media such as cornmeal dextrose agar (CMD) or white on richer media such as potato dextrose agar (PDA). Mycelium typically not obvious on CMD, conidia's typically forming within one week in compact or loose tufts in shades of green or yellow or less frequently white. Yellow pigment may be secreted into the agar, especially on PDA. A characteristic sweet or 'coconut' odor is produced by some species. Conidiophores are highly branched and thus difficult to define or measure, loosely or compactly tufted, often formed in distinct concentric rings or borne along the scant aerial hyphae. Main branches of the conidiophores produce lateral side branches that may be paired or not, the longest branches distant from the tip and often phialides arising directly from the main axis near the tip. The branches may rebranch, with the secondary branches often paired and longest secondary branches being closest to the main axis. All primary and secondary branches arise at or near 90° with respect to the main axis. The typical *Trichoderma* conidiophores with paired branches assumes a pyramidal aspect. Typically the conidiophore terminates in one or a few phialides. In some species

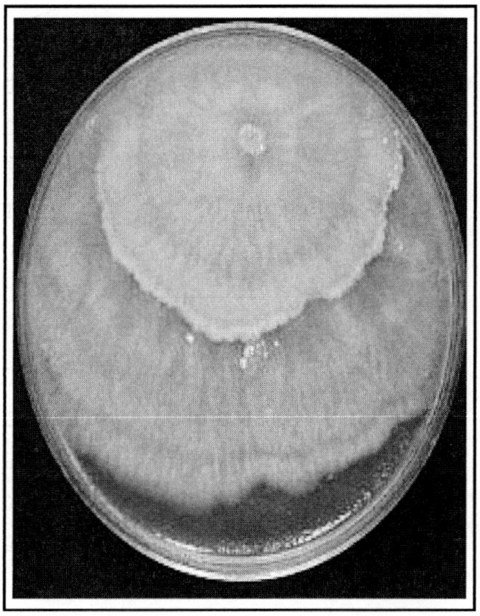

Trichoderma - fungal growth on PDA

↓

Closeup of growth

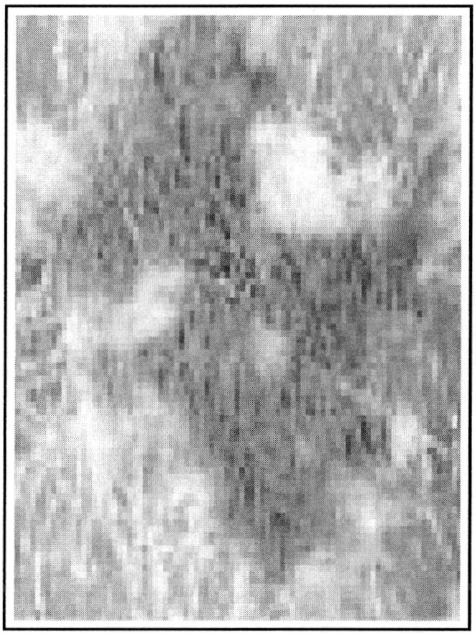

Microscopic image of *Trichoderma*

↓

Closeup of conidia

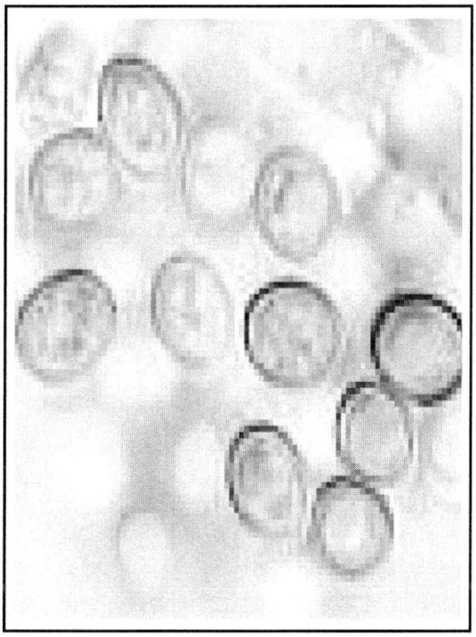

(e.g. *T. polysporum*) the main branches are terminated by long, simple or branched, hooked, straight or sinuous, septate, thin-walled, sterile or terminally fertile elongations. The main axis may be the same width as the base of the phialide or it may be much wider.

Conidia typically appear dry but in some species they may be held in drops of clear green or yellow liquid (e.g. *T. virens, T. flavofuscum*). Conidia of most species are ellipsoidal, 3-5 x 2-4 µm (L/W = > 1.3); globose conidia (L/W < 1.3) are rare. Conidia are typically smooth but tuberculate to finely warted conidia are known in a few species. Chlamydospores may be produced by all species, but not all species produce chlamydospores on CMD at 20° C within 10 days. Chlamydospores are typically unicellular subglobose and terminate short hyphae; they may also be formed within hyphal cells. Chlamydospores of some species are multicellular (e.g. *T. stromaticum*).

HABITAT

Trichoderma species are frequently isolated from forest or agricultural soils at all latitudes, species are most frequently found on bark or on decorticated wood but many species grow on bracket fungi or bird's nest fungi.

Trichoderma viride

Trichoderma viride is one such biocontrol agent, mainly used for the control of root disease of pulses and oil seeds.

Classification	Asexual (Conidial)	Sexual (ascospore)
Sub Division	Deuteromycotina	Ascomycotina
Class	Hypomycetes	Pyremomycotes
Order	Moniliales	Spaeriales
Family	Moniliaceae	Hypocreaceae
Genus	Trichoderma	Hypocrea

KEY TO IMPORTANT SPECIES OF *TRICHODERMA*

Conidiophores long and thick, often with sterile hyphal elongations side branches mostly short but thick, bearing crowded, short generally with compact tuffs of conidiophores.

1. Phialdospores Hyaline, small - *T. polysporum*
2. Phialospores green, small to large - *T. hamatum*

Conidiophores and their side branches long and slender without sterile hyphal elongations, phialides not crowded rather slender, colonies, yellowish, bright, dull to dark green, glucose or with compact conidiophore tufts.

1. Phialospores smooth-walled, ellipsoidal appears angular or oblong - *T. koningii.*
2. Phialospores, short ovoid, with length and breath more or less equal - *T. harzianum.*

Conidiophores with similar branching systems phialides irregularly disposed, often arise singly.

1. Phialospores rough-walled - *T. viride.*
2. Phialospores smaller oblong, ellipsoidal, phialides distinctly alternated at the base - *T. pseudokoningii.*

MORPHOLOGY

Colony characters

Rapidly growing colonies, mycelium floccose, woolly or aerial, conidiation effuse/tufted/compact pustules; white at first, and turns green. Chlamydospores are abundant, intercalary, or terminal on short lateral branches of vegetative hyphae, globose to ellipsoid, smooth or thick walled, color less to pale yellowish or greenish.

Conidiophores

Main axis is broad, straight, or flexuous. Primary branches arise at regular intervals and produce smaller secondary branches of branching nature, branches are more or less divergent, paired, or in vertical, repeated verticillate branching results in pyramidal structure.

Phialides

Flask shaped, slightly narrower at the base, bulged at the middle and with narrow conical or sub cylindrical neck. Phialides are disposed in verticills or in whorls beneath the septa of the conidiophore.

Phialospores

Conidia single celled, colorless/grayish/green cell wall roughened on the outer wall conidia is ellipsoid or oblong.

ISOLATION FROM SOIL

Trichoderma is isolated from the soil using *Trichoderma* selective medium developed by Elad and Chet (1983).

Collect soil samples from the field, mix well and make it into fine particles. Soil samples should be collected in root zone at 5-15 cm depth and from rhizosphere wherever possible. Ten gram of soil sample is taken, and suspended in 100 ml of sterile distilled water and stirred well to get 1:10 dilution. Transfer one ml from this dilution to 9 ml sterile water in a test tube to get 1:100 dilutions. Make serial dilutions by transferring one ml of suspension to subsequent tubes to get dilution of 1:10,000.Transfer one ml of the desired soil suspension to sterile petriplates. Pour 15 ml melted and cooled *Trichoderma* selective medium (Annx. I) in the same petriplates. Rotate the plate gently and allow to solidify. Incubate at room temperature for 5-7 days and observe for the development of fungal colonies. *Trichoderma* colonies will be white initially and turn green. Count the number of colonies developing in individual plates. Transfer the individual colonies to potato dextrose agar slants.

MASS PRODUCTION

The important species available for mass production are *T. viride and T. harzianum.* Equipments required - like fermentor, rotary mixer, auto packer, rotary shaker, laminar flow, water distillation unit, refrigerator, haemocytometer etc. are required for the production of *Trichoderma* fungi.

Major steps

Inoculation → Fermentor run → Harvesting → Blending → Drying and Packing.

Outlines

The pure mother culture of fungi is being maintained in MTCC, Agri. Universities and some ICAR institutions etc. The mother culture can be purchased from the identified sources. They have to be further sub-cultured and maintained purely for mass production by adopting standard techniques under the supervision of trained microbiologist or pathologist.

- The culture has to be mass multiplied in two levels namely (i) at primary level using shakers in flasks and (ii) secondary stage multiplication in fermentors. The important factor in this is the preparation of growing medium in which the culture is mass multiplied. For *Trichoderma* Fungi, the growing media used in the model is molasses and protein material.
- After the growing media is formulated and sterilized in fermentor, it is inoculated using the culture multiplied in the flasks.
- The molasses based culture media is continuously aerated by passing sterile air from compressors. After about 3-4 days fermentation period, the culture will be ready for packing in a carrier material.
- While the inoculated culture is gathering ready in the fermentors, the carrier material is sterilized in autoclaves and kept ready for mixing the culture. Talc powder is reported to be the commonly used carrier material for *Trichoderma* Fungi.

☆ The cultured (fungus) and sterilized carriers are mixed mechanically in a blender and the material is packed using semi automatic packing and sealing machine.

MEDIA

Molasses yeast medium **(Annexure I)** is prepared in conical flasks and sterilized at 1.1 kg/cm for 20 minutes. *T. viride* culture is inoculated by taking fungal disc from 10 day old culture and incubated for 10 days. This serves as mother culture which is added to the fermentor @ 1.5 lit/50 lit. of medium and incubated at room temperature for 10 days. The fungal biomass and broth are mixed with the powder at 1:2 ratio. The mixture is air dried and mixed with carboxy methyl cellulose (CME) @ 5 g/kg of the product. It is packed in polythene covers and used within 6 to 12 months.

QUALITY CONTROL OF PRODUCT AND PACKAGING MATERIAL

(AS PER THE SECTION 9(3) OF THE INSECTCIDES ACT 1968)

Formulation Preparation of *T. VIRIDE*

1. *Trichoderma viride* (spores & mycelia)	1.0% w/w
2. Talc	98.5% w/w
3. Carboxy methyl cellulose	0.5% w/w

TALC BASE PREPARATION

1. Form and appearance	:	Powder and dull white in appearance
2. Moisture content	:	8 per cent
3. pH	:	pH 7
4. Composition	:	*T. viride* fungal spores and mycelium 2 x 10^6 cfu/g – (1.0% w/w) Talc powder – 98.5% w/w Carboxy methylcellulose – 0.5% w/w
5. Colony forming units	:	Minimum 2 x 10^6 cfu/gram
6. Contaminants	:	*Salmonella* - absent *Shigella* - absent *Vibrio* - absent Other microbes - less than 10/g

PACKAGING AND CONTAINER CONTENT COMAPTIBILITY

Container	:	LDPE pouches
Capacity	:	100 g
Test Substance	:	*T. viride*

Description of common product

1. Physical	-	Dull white powder
2. Purity	-	100%
3. pH	-	7
4. cfu/g	-	Minimum 2 x 10^6 cfu/g on 30 days after storage at 30°C

To obtain information regarding the container compatibility of *T. viride* under different temperatures of storage for different periods. The containers to be filled with product and heat-sealed and stored at different temperatures namely 0, 10, 20, 30 and 40°C for different duration's namely 0, 15, 30, 45, 60, and 75 days.

1. Weight - Before and after the storage period, the weight was determined using analytical balance.
2. Active content

   (a) Method - Serial dilution method

   (b) Protocol

   1. One gram of the product was taken and mixed with 10 ml of sterile distilled water in a test tube. From this, serial dilutions were prepared up to 10^6 dilution was transferred aseptically to sterile Petri plates poured with *Trichoderma* special medium. It was incubated at 28 ± 2^0 C for 5 days. The colonies of *T. viride* were estimated and expressed as cfu/g of the product.

   2. *Testing of container:* The compatibility of the container was tested at different conditions namely, room temperature (30°C), cold condition (0 to 10°C) and at warm conditions (40°C).

If there is no significant changes in the physical properties like weight and biological properties (CFU/g of *T. viride*). No significant change if noticed on the container like visible cracks, weight, shape, dimension and deterioration or blooming.

Bioefficacy of *Trichoderma* spp. in India

Efficacy in cash crops

Charati *et al.,* 1998 studied efficacy of *T. viride* and *T. harzianum* in seed treatment against *R. solani* on cotton in seedling disease and also against root rot disease caused by *M. phasiolina* in cotton and found them very effective. Soil treatment of *T. viride* was found effective against *P. graminicola* in root rot disease of sugarcane, moreover, soil and spray treatment of *Streptomyces rochei, S. dayalbaghensis, A. niger, Trichoderma* spp. and *P. fluroscens* in the red rot disease of sugarcane caused by *Colletotrichum falcatum.* Seed treatment and nursery bed treatment of *T. harzianum,* was found effective against tobacco damping off disease caused by *P. apanidermatum.*

Efficacy in Pulses

Seed and soil treatment of *T. viride, T. harzianum, T. virens, G. deliquescens* and *P. fluroscens* in chickpea for management of wilt, root rot, seed rot and collar rot caused *by F. oxysporum* sp. *ciceris, S.* rolfis, *R. solani R. bataticola* and *M. phaseolina* respectively and reported very useful (Prasad and Rangeshgwan, 2000). In root rot and wilt disease of cowpea treatment of *T. viride, T.harzianum* reported satisfactory control of *M. phaseolina, F. oxysporum f.* sp. *tracheiphina.* (Ushamalini *et al.*, 1997). Kehri and Chandra (1991) studied efficacy of *T. viride* and *T.harzianum* as seed and soil treatment against *M. phaseolina* on Mungbean root rot and reported effective. In wilt disease of pigeonpea treatment of *T. viride, T.harzianum, T. koningi, T. hamatum* and *B. subtilis* reported satisfactory control of *F. udum* (Panday and Upadhyay, 1999).

Efficacy in Cereals

It is reported that application of *T. viride, T. harzianum* and *T. virens,* as soil, seed and seedling treatment control the Banded blight of rice caused by *R. solani.* In Brown spot disease of rice seed treatment of *T. viride* reported satisfactory control of *Drechslera oryzae.* Sharma and Basandrai (2000) studied efficacy of *T. viride,T .harzianum, T. viren, B. subillus, P. fluroescens* and *T. virens* as seed treatment against Karnal bunt disease of wheat caused *Neovossia indica.* Application of *T. viride, T. harzianum and T. lignorum, P. fluoescens* and *B.* subtillus as seed treatment in loose smut disease of wheat caused by *Ustilago segatum tritici* remained effective.

Efficacy in Oil Seed Crops

Seed and soil treatment of *T. viride, T. harzianum* and *B. subtilus* tested in groundnut against collar rot , root rot and crown rot caused *by A. flavus, S. rolfis* and *A. niger* (Thirumala and Sitaramaiah,2000; Dubey, 2000). In Damping off disease of mustard seed and soil treatment of *T. viride, T.harzianum* and *Latisaria arvalis* reported satisfactory control of *P. aphanidermatum* (Chattopadhyay *et al.*, 2002) In root rot disease of soyabean treatment of *T. viride* and *T. harzianum* reported satisfactory control of *M.phaseolina.* Efficacy of *T. hamatum, T.harzianum, P.* flurorescens and *P. putida* as seed treatment on sunflower root rot which is caused by *S. rolfis, R. solani* and *S. sclerotium* remained very successful.

Efficacy in Vegetables

Seed treatment of *T. harzianum, T. viride* and *T. koningii* on cabbage damping off disease caused by *R. solani* found effective. Good effectiveness seen for *G. macrocarpus* and *T. harzianum* in soil treatment in chilli crop for control of root rot disease cause by *S .rolfis*. Seed and soil treatment of *T. harzianum, T. viride* and *T. koningii* in egg plant against wilt and damping off caused by *F. solani* and *P. aphanidrmatum* well managed.

APPLICATION AND DOSAGE

Talc based formulations of the fungal antagonists are applied at the rate of 4gm per kg of seed for controlling soil borne plant diseases. Mix the powder with sufficient quantity of water to make slurry for treating seed before sowing.

Soil Application

Trichoderma spp. suppress the activity of soil borne fungal pathogens, especially *Rhizoctonia* and *Pythium* and protect transplanted seedlings by colonizing their roots.

Seed Treatment

Seed treatment is an alternative approach to introduce *Trichoderma* spp. into the soil. This method requires smaller amounts of biological material than soil treatment. Unlike chemical fungicides, *Trichoderma* spp. provides long-term protection without any adverse side effects.

ADVANTAGES

- Can be used along with organic manure.
- *Trichoderma* spp. are also known to suppress plant parasitic nematodes (root-knot nematodes) in addition to other diseases.
- Lower cost and longer efficacy than conventional fungicides.
- Does not lead to development of resistance in plant pathogens.
- No phyto-toxic effects and ecofriendly.
- Minimizes losses and cost of production and increases yield and profit.

—•—

Chapter 16

Paecilomyces lilacinus A Bioagent of Root-Knot Nematodes and Promot Plant Growth

Phyto-nematodes belonging to genus *Meloidogyne* (Nematoda: Tylenchida) causing root-knot disease by producing galls on the roots of wide varieties of agricultural crops are the important global pests. Amongst the 98 species of *Meloidogyne* reported so far world over, twelve species with few pathotypes are occurring in India. RKNs are the key nematodes widely distributed throughout the country causing discernible yield losses both quantitatively as well as qualitatively in almost all important agricultural and horticultural crops attacking more than 3000 plants. Biological control of root-knot nematodes is cost effective and ecofriendly new approach in India. It includes use of bacterial, fungal and other parasites and predators for effective and economic suppression of root-knot disease. But, mainly the progress is being made for effective utilization of former two groups of bioagents. Suitable mass production techniques and formulation preparations are evolved for further increasing effectiveness of these bioagents. Compatibility of bioagents with nematicide and other pesticides is an added advantage exploring best possibility for using them in Integrated Nematode Management (INM) to reduce pesticidal load in ecosystem to preserve our environment at long run.

Root-knot nematodes (RKN) are the species of nematode genera *Meloidogyne* producing galls on the roots of wide varieties of cultivated plants providing food, fiber and timber. They are also considered as one of the important global pests. Twelve species of RKN with pathotypes are reported from India attacking almost all agricultural crops amongst 63 species known in the world. However, RKN are key nematodes widely distributed throughout the country causing discernible crop yield losses (Table 16.1) both quantitatively as well as qualitatively in various crops. Presently for management of RKN, only one insecticide based nematicide carbofuran is available to the growers in Indian market. Having only this nematicide available and its soil and environmental pollution as well as health hazardous effects, biological control of nematodes is an important alternative tool for effective and economic management of RKN.

INTERNATIONAL STATUS

For biological control of RKN, more than 100 bioagents belonging to fungi, bacteria, viruses, nematodes, protozoan, etc. are reported world over in last 5 decades (Table 16.2). Of which few promising fungal agents were identified and now commercialized and marketed for farmer's use in last decade. The Royal 300 and 350 products based on *Arthrobotrys* spp. in U.K., *Paecilomyces lilacinus* based product Biocon in Philippines (Asia) and Peru (South America), are in use. Recently, a *Bacillus* antagonist of RKN is identified and developed by Ecogen Inc., which will be shortly available in the market of Israel. This shows an adaptability and utility of biocontrol agents for RKN management in the world.

NATIONAL STATUS

In India, research work on biological control of nematodes was started in PL-480 projects in late 1960's using imported cultures of nematode trapping fungi, *Dactylaria* and *Dactylella* at IARI, New Delhi; AMU, Aligarh and GAU, Anand. Systematic research efforts were made by isolation of 7 native nematophilic fungi at IARI, New Delhi.

Nematophilic Bacteria

Gokte and Swarup tested five bacterial isolates against *M. incognita* and reported *Bacillus subtilis* and *P. pumilis* giving 100% larval mortality at 2.0 O.D. concentration. *Pasteuria penetrans* against *M. incognita* J2 was studied in laboratory as well as on tomato plants and reported 92% spore attachment on larvae with 65 and 51 % reduction of nematode penetration in host root and gall formation respectively @ 3.5 g spores/kg sand in pots. Reduction of 94% *M. javanica* population by *P. penetrans* at 1000 mg/kg soil on tomato in pots was reported, further studied the effect of *P. penetrans* and carbofuran alone as well as in combinations and reported 69 and 63% reduction in gall formation by RKN in alone and combined treatments respectively (Table 16.3). Though, the bacterium is highly infective but due to its obligate parasitic nature, *in vitro* mass production and use is limited in the country. A broad-spectrum bioagent, *P. fluorescence,* generally useful for soil born diseases is also reported beneficial in nematode management.

Nematophilic Fungi

Among several bioagents, nematophilic fungi are well studied and useful. Several predaceous and parasitic fungi are reported attacking RKN in India. Culture filtrates of *Arthrobotrys, Dactylaria* and *Curvularia* effect on hatching of *M. incognita* eggs showed that 100 % concentrated filtrate of *Arthrobotrys* allowed only 3.8 larval hatching compared to 45 in water. Different *P. lilacinus* doses against *M. incognita* on brinjal in pots was tried and reported 75 g dose to be most effective giving 91% nematode control (Table 16.4). *P. lilacinus* followed by *M. incognita* in sequential application realized maximum yield and minimum nematode infection in okra. Among several fungal bioagents, predaceous and parasitic fungi *Arthrobotrys, Dactylaria* and *Curvularia* are reported attacking RKN in India since 1960s. Opportunistic fungal bioagents have easy and economic *in vitro* mass production technology, hence more popularized for biological control of root-knot disease in the country. Biological control of root-knot disease by nematode egg parasitizing fungus, *P. lilacinus* in groundnut and cotton has been successfully demonstrated in Gujarat (Table 16.5). Reduction of root-knot nematode by 34.7 per cent in groundnut and 28.7 % in cotton increasing 19.1 and 10.7 per cent crop yield, respectively over control was reported at fungus dose 25 kg ha^{-1} (spore dust/granules having 10^9 conidia/g based on rice grain substrate carrier). In double whammy approach, *P. lilacinus* with nematicide and soil organic amendments gave better control of root-knot disease in cotton, chickpea, and banana as well as tomato nursery over the last decade. Two products based on these fungal bioagent are also marketed recently.

Multipurpose utility of bioagents is also successfully achieved in India, entomopathogenic fungi viz, *Beauveria brongniartii* and *Metarhizium anisopliae* were proved to be parasitic on eggs of root-

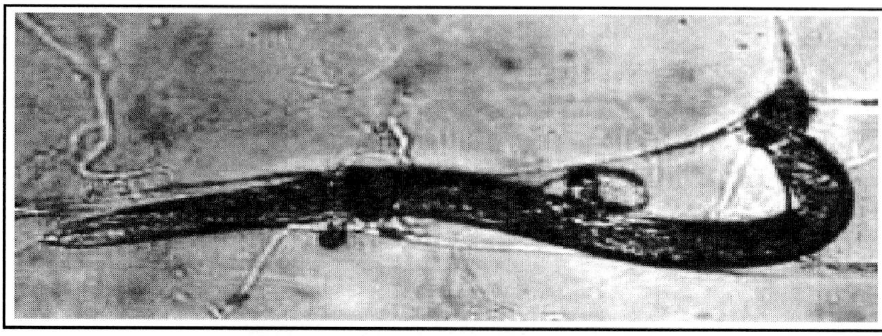

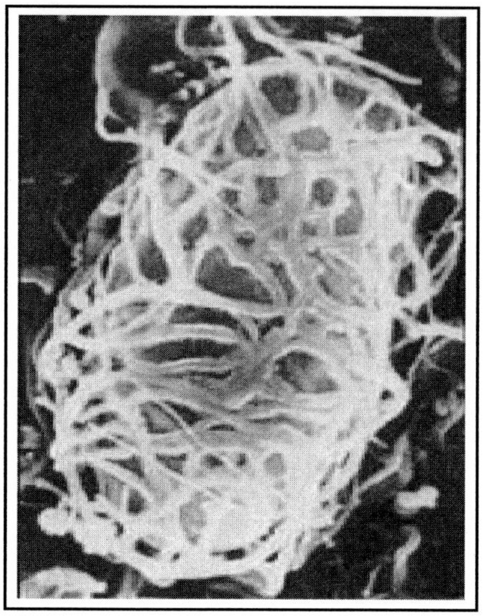

Fungus infected nematode egg

knot nematodes. Similarly, *T. harzianum,* a fungal antagonist is also reported effective against root-knot nematodes. History, development and progress made so far on bionematicides against root-knot nematodes attacking agricultural crops in India is documented. Study conducted on bioagents, *P. penetrans* and *P. lilacinus* alone and in combinations against *M. javanica* on tomato indicated 66 % reduction in nematode population in combined treatment. Castor leaves as organic amendments were evaluated with *P. lilacinus* alone and in different sequences against *M. Javanica* and found application of castor leaves and fungus 15 day prior to nematode application to be most effective giving maximal yield with minimal gall index. Application of *P. lilacinus* at 20 g/m2 with carbofuran at 2 g/m2 effectively reduced *M. arenaria* gall index on brinjal seedlings.

MASS PRODUCTION OF BIOAGENTS

Among bacterial pathogens, only *P. penetrans* is reported highly pathogenic, which is an obligate bacterium and can't be cultured on artificial media in vitro. Hence, its mass production and use against RKN is very limited. In fungal agents, *P. lilacinus* is found most virulent and easy to mass culture by solid substrate fermentation on broken grains of wheat or rice world over. Hence, it is well adopted by scientists and companies in developed countries. In India, maximum growth and sporulation of *P. lilacinus* on mungbean husk followed by cotton and sesamum oil cakes amongst various agricultural byproducts tested. GAU reported karnaj and neem cakes to support good growth and sporulation of *P. lilacinus*. They also found broken wheat and pearl millet grains as best substrates for mass production of fungus.

Mass Culturing and Maintenance of Fungus in Laboratory

Stock culture maintained on selective media used for the routine maintenance of isolate and for production of conidia potato dextrose broth/agar is to be used **(Annexure I).**

Fungus stock culture is usually accomplished on selective agar media using glass Petri plates in the laboratory. For mass production, PDA plates (50) are inoculated under sterile conditions @ 10^4 spores/ plate from stock fungal culture. But, before that dodine is added by the time of pouring of media in to plates. It is observed that even 300 ìg/ml of dodine-inhibited isolation of some isolates of *M. anisopliae*. They found that a reduced concentration of dodine (10 µg/ml) increased the efficacy of recovery of desired fungus and inhibited common fungi. PDA dishes are incubated at 25-27°C under B.O.D. incubator. To reduce the dehydration of the media, petri dishes are sealed with para film. Within 8-12 days, the culture is sporulated and the conidia are harvested in sterile D/W 5ml/plate by scrapping from the surface using a sterile glass spreader.

MASS PRODUCTION OF *P. lilacinus* ON SOLID SUBSTRATE

Mass multiplication of fungus is carried out on broken rice grains in autoclavable HDPE bags (High Density Poly Ethylene) of 20 x 25 cm size. Broken rice grains 100g are soaked in 90ml water and bags are sealed. Bags with grains are autoclaved at 121°c and 15 lbs pressure for 30min. On cooling one corner of bag is cut and 5ml spore suspension prepared in sterile distilled water having spore count 10^8 /ml is inoculated in sterile condition in laminar air flow chamber. Bags are than incubated in BOD incubator of 27 + 2°C for 15 days. After 15 days spores along with grain carrier are harvested from bags and dried in natural sunlight in net house for 2 - 3 days. Dry pallets are crushed in household mixture to get fine spore powder based on substrate. Spore dust thus mass produce is having 10^9 spores/g strength and used for management of root-knot nematodes in different crop fields.

RESEARCH AT AAU

Laboratory, pot and field studies on bioefficacy of *P. lilacinus* have been carried out at Dept. of Nematology, GAU, Anand. Vyas *et al.,* (1996) have reported 64 % mortality of *M. incognita* eggs at 5 x 10^7 spore dose in laboratory bioassay. Bioefficacy of *P. lilacinus* against *M. javanica* (patho.2) on groundnut in pots was tested and reported 3 to 5% (w/w) fungus application to be effective for nematode management but seed dressing by fungal spores proved ineffective. *In vitro* compatibility of *P. lilacinus* with various nematicides indicated that fungus is well compatible with phenamiphos, sebufos, carbofuran, phorate, benfurocarb and terbufos at field recommended, double and half doses. Integrated management of RKN on chickpea with *P. lilacinus,* carbofuran & phenamiphos and their combinations were studied and found significant reduction in RKI in treatment of fungus with phenamiphos and thereby gave higher yield over control. Integrated management of root-knot nematode in tomato nursery 4 yrs pooled results indicated that soil solarization with 25 LLDPE clear film in hot summer for 15 days is mot effective technique. Soil application of *P. lilacinus* pellets @ 25 kg/ha neem cake 1 t/ha and carbofuran 1 kg a.i/ha remained less effective and at par to each other but better then control. However application of fungus followed by soil solarization found more effective.

Efficacy of *P. lilacinus* was also evaluated on groundnut at farmer's field at Lalpur & Kabhai na muvada villages (Kapadwanj taluka). Three years pooled results indicated that fungus application @ 50 kg/ ha gave maximum pod yield with reduction in RKI. Lower dose of 25 kg/ha exhibited almost similar results as carbofuran treatment @ 1 kg/ha but better than check. Integrated management of RKN on cotton Hy.8 was tried with *P. lilacinus,* Sincocin 98 AG, carbofuran and phenamiphos alone as well as in combinations, three years pooled results revealed that phenamiphos and fungus alone as well as in combination gave lowest RKI and higher yield compared to other treatments.

GRANULAR FORMULTION OF FUNGUS *P. LILACINUS*

Moreover the muscardine fungi, *M. anisopliae* and *B. brongniartii* (powerful bioagents for biocontrol of soil insects) are parasitic to *M. incognita* eggs was reported for the first time in the country. These two fungi were also tested alone and in combinations with *P. lilacinus* against *M. javanica* (patho.2) and white grubs attacking groundnut in pots. Muscardine fungi were found to manage white grubs upto 85 % with simultaneously managing RKN. But *P. lilacinus* could not manage white grubs. Under field conditions also these muscardine fungi with pesticides effectively reduced *M. javanica* ptho.2 on groundnut.

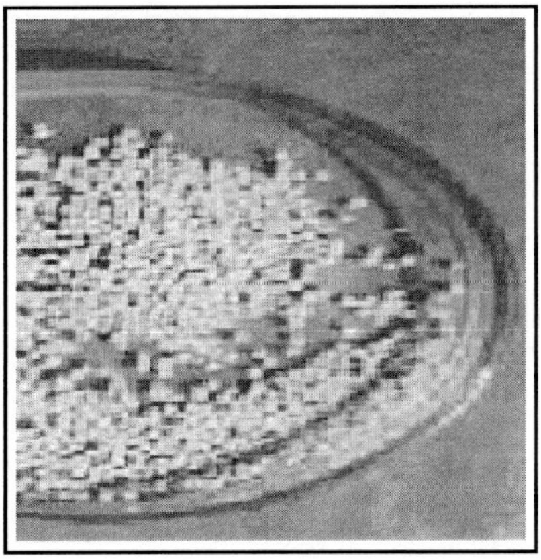

(Granual formulation) of *P. lilacinus*

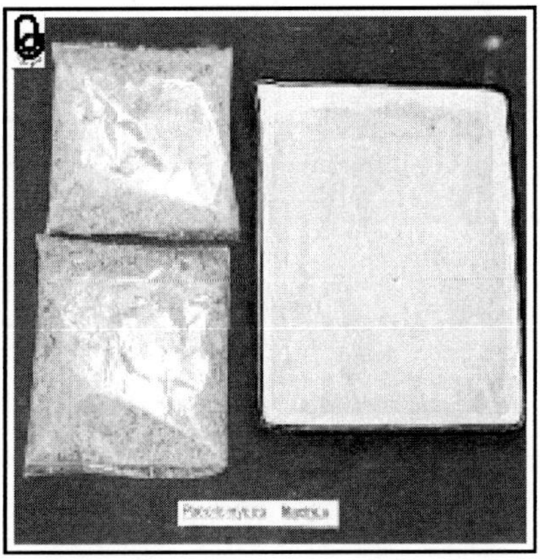

P. lilacinus growth in HDPE bags and spore dust

Future Thrusts

- Due to chemical nematicide crisis in India, there is an urgent need to promote environmental friendly bio- nematicides in the country. Moreover recent Government policies also favor biopesticides.
- Industrial fermentation technique for fungal bioagent *P. lilacinus* needs to be developed to launch a bionematicide in the market.
- Large-scale field demonstration at farmer's fields is required to increase awareness and adoption of bioagents for RKN management.
- Use of effective bioagents should be accelerated in integrated nematode management and economics needs to be worked out.

Table 16.1. Yield losses in different crops due to root-knot Nematodes in India.

Crop	% Loss	Reported by and Year (Reference)
Vegetables		
Brinjal	35-75	Srivastava, 1969
Chilli	70	Sen, 1958
Okra	90	Bhatti & Jain, 197
Tomato	20-75	Gaur, 1972
Pulses		
Blackgram	40-46	Kalita & Phyken, 1989
Chickpea	28-40	Daraker & Jagdale, 1989
Cowpea	28	Reddy & Singh, 1981
Mungbean	8-82	Gaur, 1972
Pea	19-20	Reddy, 1985
Species		
Blackpepper	50	Koshy *et al,* 1979
Ginger	74	Sukumaran & Sundarrajan, 1986
Cash crops & others		
Banana	15	Patel *et al.*,1999
Cotton	10-15	Palanisamy & Bala Subramanyan, 1982
Groudnnut	13-59	Vaishnav *et al.*, 1985
Rice	21-64	Kumari & Kuriym, 1991
Tobacco	50-51	Patel *et al.*, 1986
Wheat	35	Patel & Patel, 1998

Table 16.2. Key bioagents of root-knot nematodes (RKNs)

Name	Mode	Action	Adoptability/ Product
1. Fungi			
Arthrobotrys spp.	Predaceous	Sticky network, Secret toxic substance	Wide use, products Royal 300 & 350 in UK

Contd...

Table 16.2. Contd...

Dactylella spp.	Predaceous	Constructing ring	Common use
Dactylaria spp.	Predaceous	Sticky knobs	Common use, Product Prospernema, Circle one Inc, USA
Monacrosporium spp.	Predaceous	Adhesive branches	Common use
Harposporium sp. *Hirsutella* spp.	Endozoic (obligate)	Adhesive spores, attach and penetrate larvae, Fill up body with fungus mycelium	Rare use
Gliocladium virens *Paecilomyces lilacinus* *Metarhizium anisopliae* *Verticillium-chlamydosporium* *Myrothecium* sp.	Opportunistic and parasitic	Colonize egg masses, attack female reproductive structures, action of enzyme and toxin help in rapid killing of host	Wide use in UK, USA & Australia; *Paecilomyces*, product Biocon, Philippines, Paru; *Metarhizium*, product Biogreen, Australia; DiTera, Abbott, USA.
Vesicular Arbuscular Mycorrhizae (VAM)	Antagonistic	Colonized roots Restrict nematode entry, high amino acids and $P0_4$ in the colonized plant roots also helps in repelling J2	Common use in several countries, product Prospernema, Circle one Inc, USA
2. BACTERIA			
Pasteuria (Bacillus penetrans)	Obligate	Spore, attach on Cuticle and subsequently hyphae penetration and proliferation in larval body, forming colony having million of spores / J2	Wide use, Product not available

Contd...

Table 16.2. Contd...

B. thuringiensis	Opportunistic	Endotoxin kills J2	Restricted use
Bacillus sp., *B. subtilis,* *B. pumilis,* *Clostridium sp.*	Opportunistic	Invade in larvae and kill, metabolites produced are also toxic to J2	Common use, *Bacillus* based product Ecogen Inc, Israel
3. PROTOZOA			
Legerella sp. *Leptomonas* sp. *Theraromyxa* sp.	Parasite Parasite Predator	*Coccidian* parasite of larvae Engulf larvae	Rare use
4. NEMATODES			
Mononchida spp. *Diplogasterida* spp. *Dorylaimida* spp.	Predator Predator Predator	Using sucking teeth feed on juveniles	Rare use
5. MISCELLANEOUS AGENTS			
Polyhedral virus Rickettsia Mites	Obligate Obligate Opportunistic	Infection reduce life of nematode Feed on larvae attaching on cuticle	Rare use
Earthworm	Predator	Engulf larvae along with soil	Rare use
Soil arthropods (*Collembdans*)	Predator	Caught larvae and ingest	Rare use

Table 16.3. Effect of nursery treatment with P. *penetrans* and carbofuran on gall formation and growth of tomato cv. Pusa ruby.

Treatment	Fresh wt., g		Length, cm		No. of galls/plant
	Shoot	Root	Shoot	Root	
CK (3000 J2)	0.60 a	0.10 a	3.40 a	2.65 a	12.17 a
CK (Un inoculated.)	2.00 b	0.55 b	6.57 b	7.53 b	—
Pp (30 lake spores)	1.42 c	0.40 c	5.50 e	5.50 e	1.73 b (-69.35 %)
CAR (1.5kg/ha)	1.10 d	0.33 d	4.08 d	3.50 d	5.96 c (-51.03 %)
Pp + CAR (Half doses)	1.27 cd	0.37 e	4.84 e	4.63 e	4.50 b (-63.02 %)
C.D. 0.05	0.18	0.18	0.26	0.32	0.82

Figures in parentheses are percent decrease (-) over control.

Pp- *Pasteuria penetrans;* CAR- Carbofuran; CK- Control.

Table 16.4. Control of *M. arenaria* in brinjal cv. MDV 1 nursery.

Pooled (Two years)

TRT	Fresh wt., g (n=25)	No. of galls/plant
CAR 1 g/m^2	116.7 c	18.5 b
CAR 2 g/ m^2	125.7 b	9.1 c
CAR 1 g + Pl 20 g/ m^2	135.3 a	9.0 c
CAR 2 g + Pl 30 g/ m^2	121.8 e	4.5 d
Pl 20 g / m^2	121.8 e	18.9 b
CK	109.0 d	32.4 a

Figures indicating common letters do not differ significantly from each other according to DNMRT.

Pl- *Paecilomyces lilacinus;* CAR- Carbofuran; CK- Control.

Table 16.5. Bio-efficacy of *Paecilomyces lilacinus* for the management of root-knot disease on cotton

Pooled (1996-98)

Treatment	Plant height, cm.	Yield, kg/ha	RKI (0-5)*
PL	110.03bc	2034.02 cd (+10.7)	2.21 bcd (-28.7)
PHEN	110.89ab	2182.13 abc (+18.7)	1.84 e (-40.6)
CAR	110.51b	2056.03cd (+11.9)	2.15bcd (-30.6)
SINC	109.31bc	1915.29de (+4.2)	2.86ab (-7.7)
PL + PHEN	112.87a	2268.21a (+23.4)	1.51f (-51.2)
PL + CAR	112.02ab	2191.56ab (+19.2)	2.02de (-34.8)
PL + SINC	110.35bc	1915.42de (+4.2)	2.50bc (-19.3)
CON	108.49c	1837.77e	3.10a
S.Em	0.61	56.65	0.11
C.V.%	1.41	10.05	16.79

* (O = Free, 5 = Maximum disease intensity)

—•—

Chapter 17

Bacillus thuringiensis A Key Biopesticide

Biopesticides are potential alternative to chemical pesticides and widely recommended as a component in Integrated Pest Management (IPM). These eco-friendly natural enemies of insects are derived from three sources *viz.,* microbial, beneficial arthropods and botanicals. Insect pathogenic microorganisms belonging to bacteria, fungi, viruses, protozoan, EPNs etc. are considered as microbial pesticides. The most commonly used microbial agents to control noxious insects world over is *B. thuringiensis (B.t.).* The efficacy of *B.t.* is quite impressive than other biopesticide and easily available in market. Amongst all insect pathogenic microbes, *B.t.* has received prime attention and the product based on this bacterium is widely demonstrated among all the biological insecticides.

B.t. is a gram positive, aerobic, sporulating bacterium that synthesizes crystalline proteins during later stage of growth as secondary metabolites which is highly toxic to agriculturally important pest especially caterpillars belonging to lepidopterans at very low concentration. The outstanding feature of *B.t.* is the production of protenaceous delta (ä) endotoxin (crystal) that accounts up to 30 % of total protein content of the bacterium. The crystal contains insecticidal proteins that may vary in type, quantity as well as toxicity against different groups of insects depending on the bacterial strain and place of isolation. *B.t.* ä endotoxin (cry proteins) has acquired acceptability as eco-friendly biopesticides because of its specificity towards insects having alkaline gut and by enlarge safer to higher animals. *B.t.* is under extensive use for control of insects in agriculture, horticulture, forestry and also for mosquito control over past three decades.

HISTORY

The discovery and historical development of *B.t.* is quite old and interesting, key events are highlighted in Table 17.1.

Table 17.1. Key events on *B. thuringiensis* in 20th century.

1902	Japanese bacteriologist, Ashimata, Shigetome, isolated "Sotto bacillus".
1915	Named as *Bacillus thuringiensis* by Berliner.
1927	*B.t.* var. *kurstaki (B.t.k.)* was introduced in Germany targeting lepidopteran insects
1931	Successful field tests against European Corn borer by Ituiz
1938	Commercial *B.t.k.* product "Sporeine" released in France
1940	Sandoz company started large scale production of *B.t.k.* based product "Thuricide"
1951	Successful field tests against alfalfa caterpillar in California by Steinhaus
1952	American firms began research on *B.t.*
1953	Hannay reported the crystal
1954	Angus demonstrated the effect of the δ-endotoxin
1958	First commercial product available in U.S.A.
1966	Dulmage isolated HD-1 strain of *B.t.* var. *kurstaki*
1971	High potency strains in commercial production
1973	Agreement on bioassay for standardization against E-61
1980	WHO reported *B.t.* var. *kurstaki* non-pathogenic to mammals
1981	Scientist Ernest Schnepf and Helen Whiteley clone a *B.t.* var. *kurstaki* gene
1985	Estimated Global sales of *B.t.* insecticides around $ 50 million
1990	*B.t.* protected transgenic crops field tested
1995	Boll Gard (*B.t.*) launched in U.S.A.
1996	Angard (*B.t.*) cotton launched in Australia

MODE OF ACTION

The crystal protein of *B.t.* is effective to insects having specific gut pH (alkaline) and /or the specific gut membrane structures required for binding of the toxin. The protein toxin binds to specific receptors present in the midgut epithelial membrane and disrupts membrane's ion exchange potential and osmotic equilibrium causing cell lyses and finally this damage leads to gut paralysis and ultimately cause death of the susceptible host. Infected insect stop feeding and die due to the combined effect like starvation and tissue damage. *B.t.* spores have also capacity to cause disease (septicemia) in the insect by invading in heamocoel system.

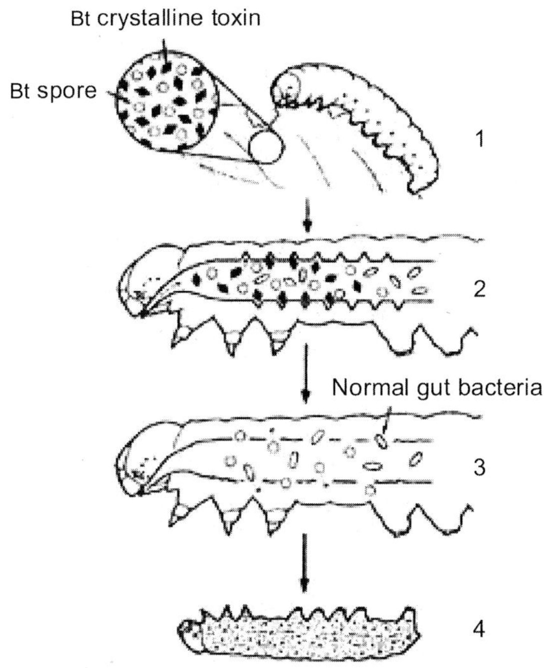

1) Caterpillar consumes foliage treated with Bt (spores and crystalline toxin).
2) Within minutes the toxin binds to specific receptors in the gut wall, and the caterpillar stops leading.
3) Within hours, the gut wall breaks down, allowing spores and normal gut bacteria to enter the body cavity; the toxin dissolves.
4) In 1-2 days, the caterpillar dies from septicemia as spores and gut bacteria proliferate in its blood.

Action of *Bacillus thuringiensis* var. *kurstaki* on caterpillars

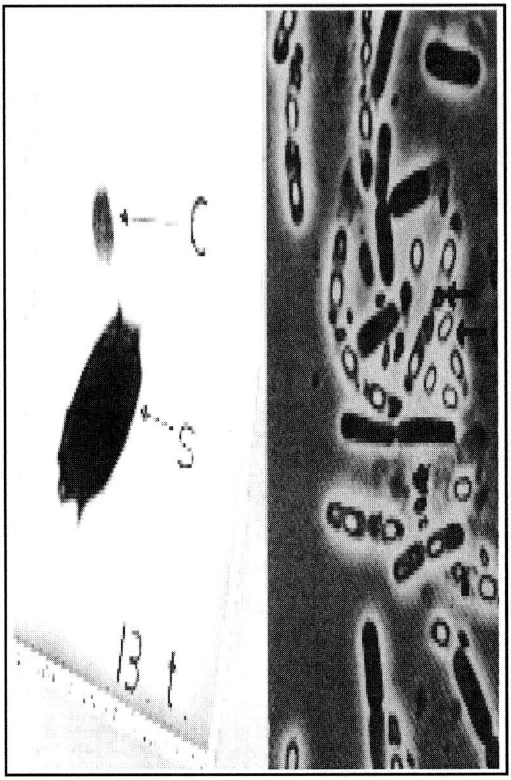

Fig. 17.1. Action of *B. thuringiensis* on caterpillars with close-up of spore and crystal toxin.

Symptoms

Larvae infected by *B.t.* become inactive, stop feeding and may regurgitate or have watery excrement. The head capsule may appear to be larger compared to body size. The larva becomes flaccid and dies, usually within 2 days. A typical putrefied smell comes out from dead cadavers. The body content turns brownish-black as they decompose, while in case of other bacteria it may turn red or yellow.

B.t. was first found to affect a range of lepidopteran insects, the major agricultural pests. Subsequently, discovery of newer strains expanded its host range. The host range of *B.t.* is constantly widening with the large amount of commercial interest and at present, extensive screening programmes are in progress world over. Although most of the *B.t.* strains are toxic to lepidoptera, few strains are known to be toxic against Coleopterans (root–grubs), Dipterans larvae especially mosquitoes, lice, mites and phytonematodes too (Table 17.2).

Table 17.2. Host range of *B. thuringiensis*

Order	**Susceptible families Most of the lepidopteran families susceptible**	
Insecta		
Lepidopterans	Sphingidae	Hawk moths
	Pieridae	Cabbage worms
	Lymantridae	Tussock moths
	Tortricidae	Leaf roller moth
	Noctuidae	Cutworms
Dipterans	Culcidae	Mosquitoes
	Simuliidae	Black flies
	Anisopodidae	Gnats
	Chironomidae	Nudges
	Psychodiae	Moth flies
	Tipulidae	Crane flies
Coleopteran	Cyrysomelidae	Leaf beetles
Phthiraptera	Philopteridae	Bird lice
	Treichodectidae	Mammalian lice
Arachnida	Dermanyssidae	Animal mites
Acari	Tetranichidae	Plant mites
Nematoda	Trichostrongylidae	Animal parasitic
Strongylida	Tylenchidae	Plant parasitic

MARKET STATUS

Among all the biopesticide in use (>5 %), *B.t.* based biopesticides are widely used in the world (up to 2 %). The total sale of spray able *B.t.* products was estimated to be 145 million in 1999, which was about 95% compared to other biopesticides. Insect resistances transgenic plants have been developed using Cry (crystal) genes of *B.t.* are further advantageous.

Table 17.3. Commercial products of *B. thuringiensis* (World market)

Trade Name	Manufacture	*Bt* strain	Used on crops & pests
Lepidopterans			
Agree	Ciba-Geigy	*Kurstaki HD-1 + aizawai*	Vegetable *Spodoptera litura*
Biobit	Novo-Nordisk	*Kurstaki HD-1*	Vegetable pests
Certan	Sandoz	*aizawai*	*Galleria mellonella*
Condor	Ecogen	*Kurstaki HD-1*	Forest pests
Cutlass	Ecogen	*Kurstaki HD-1*	Vegetable/crop pests
Cybout	Cynamid	*Kurstaki HD-1*	Vegetable/crop pests
Dipel	Abbot	*Kurstaki HD-1*	Vegetable/crop pests
Foray	Novo-Nordisk	*Kurstaki HD-1*	Forest pests
MVP	Mycogen	*pseudomonas*	Vegetable pests
Thuricide	Sandoz	*Kurstaki HD-1*	Vegetable/crop pests
Coleopteran			
Diterra	Abbot	*Tenebrionis*	Vegetable pests
Foil	Ecogen	*Tenebrionis*	*Ostrinia nubilalis* (Hubner)
M-one	Mycogen	*Tenebrionis*	*Leptinotersa decemilneata* (Say)
M-Trak	Mycogen	*Tenebrionis*	Vegetable pests
Novodor	Novo-Nordisk	*Tenebrionis*	Vegetable pests
Trident	Sandoz	*Tenebrionis*	Vegetable pests
Trident-II	Sandoz	*Tenebrionis*	Vegetable pests
Dipterans			
Acrobe	Cynamid	*Israelensis*	Medical pests
Skeetal	Novo-Nordisk	*Israelensis*	Medical pests
Teknar	Zoecon/ Sandoz	*Israelensis*	Medical pests
Vectobac	Abbot	*Israelensis*	Medical pests

MAINTENANCE AND SMALL SCALE MASS PRODUCTION

Media (Composition and Preparation)

All the ingredients of yeast Extract Mannitol Polymyxin - B (YEMA) medium (Annexure I) listed in part A are thoroughly mixed in D/W in a conical flask. Then, the media is to be autoclaved at 10 lbs pressure for 15 minutes. The egg yolk emulsion prepared from a fresh hen egg separately. The egg is surface sterilized with rectified spirit and gently broken with the help of sterile glass rod and the white portion of egg is to be discarded and yellow part, the yolk is carefully collected and diluted in sterile D/W to make 10 per cent suspension under aseptic conditions. Dissolved Polymixin-B-Sulphate powder (100 mg) in 10 ml of sterile D/W. Thus, the ingredients listed in part B are made ready as stock solution in sterile distilled water and kept in refrigerator at 5-8° C. Stock solution of egg yolk emulsion (10 ml) and Polymixin-B-suspension (1 ml) required to be added into the autoclaved medium (100 ml) with the help of sterile pipette. Then, medium is poured into sterile petriplates in laminar flow chamber.

The inoculated petriplates are incubated in B.O.D. incubator at 30 $\pm$ 2° C temperature for optimum growth. The growth characteristics of the bacterial colonies are recorded after 48 hrs. of incubation. The isolated bacterial spores are observed under phase contrast microscope for characterization and identification. The pure bacterial culture of *B. thuringiensis* is further maintained on the Nutrient Agar slants and is to be mass multiplied in Luria Broth for laboratory testing against insects. To harvest spore and crystal from broth it is to be centrifuged @ 10,000 rpm for 15 minutes. The pellet is re-suspended in sterile D/W and the respective suspensions are required to be spreaded over the host leaves for testing Koch's postulates against the targeted hosts in the laboratory.

The ingredients listed for Nutrient Agar medium (Annexure I) are thoroughly mixed in D/W in a conical flask. The medium is autoclaved at 10 lbs pressure for 15 minutes. The autoclaved N-agar medium is poured into sterile petriplates and inoculated with bacterial culture under laminar airflow.

The ingredients listed for Luria Broth (Annexure I) are thoroughly mixed in D/W in a conical flask and was autoclaved at 10 lbs pressure for 15 minutes. After cooling, *B.t.* cultures were individually inoculated into flasks and were kept on mechanical shaker and incubated for 24 hrs. at 200 rpm. Broth was centrifuged at 10,000 rpm for 15 minutes for spreading spores and crystal, from the broth. After centrifugation, the pellet of spore-crystal mixture was suspended in cold sterile D/W and used for further studies.

ISOLATION OF NATIVE *B.thuriengiensis*

Collection of Sample

Dead insect larvae should be periodically collected from fields and/or soil samples were collected from different crop fields for search of *B.t.* The samples were taken with the help of a spade and top 15 cm soil collected in the individual polythene bags and samples were labeled as per the respective crops. One hundred gram composite soil sample of each crop is to be suspended in 100 ml sterile D/W in conical flask and mixed thoroughly on mechanical stirrer for 5 minutes. Each sample (0.1 ml) is taken with the help of sterile pipette into a test tube containing 9.9 ml of sterile D/W to prepare 10^{-2} dilution and serially diluted further upto 10^{-6} dilutions. Finally, 0.1 ml 10^{-6} diluted sample of each crop was spreaded individually on the selective media, Yeast Extract Mannitol Agar with Polymixin-B for selective isolation of *B.t.* The procedure was carried out in laminar airflow chamber to avoid the contamination. The inoculated petriplates were incubated in B.O.D. incubator at 30 $\pm$ 2° C temperature for optimum growth. The growth characteristics of the bacterial colonies were recorded after 48 hrs. of incubation. The isolated bacterial spores were observed under phase contrast microscope for characterization and identification.

MAJOR STEPS IN FERMENTATION PRODUCTION PROCESS

Inoculation → Fermentor → Harvesting → Blending → Drying and Packing.

Fermentation medium

Liquid media		Semi solid media	
Contents	**g/L**	**Contents**	**g/L**
Soybean meal	15.0	Wheat bran	545
Dextrose	5.0	Expanded per liter	380
Corn starch	5.0	Soybean meal	62
$Mg\,SO_4 \cdot 7\,H_2O$	0.3	Dextrose	36

Contd...

$Fe\,SO_4.7\,H_2O$	0.02	Lime	3.6
$Zn\,SO_4.7\,H_2O$	0.02	NaCl	0.9
$Ca\,CO_3$	1.0	$CaCl_2$	0.29
		Pure water	160 ml

Note: Rest of the procedure and precautions, packaging etc. remains same as discussed for *T. viride,* Chap. 8.1

AAU STUDIES ON NATIVE *BT*

Some studies were also carried out on use of native *Bacillus thuringiensis* in last five years (2000-05) at AAU, Anand. The results of various experiments carried out on isolation of *B. thuringiensis* and bio-efficacy against castor pests viz., *Spodoptera litura* and *Achea janata* revealed that four native *Bt* strains were isolated at Anand in year 2000 and all were found very effective on both the pests, but isolate GAU CT–2 gave highest mortality of *A. janata* (83.33%) and *S. litura* (88.89%) respectively with LC_{50} of 1.8 to 5.6 x 10^8 spores/ml. Other native isolates were further studied and research findings on *B. thuringiensis* isolates (GAU TOM-1, GAU TOB-1 and GAU CT-2) against second instar larvae of *M. testulalis* in laboratory and pot testing produce considerably good mortality (66 to 80 %). In cowpea field these native isolates of *B.t.* performed effective control of *M. testulalis,* exhibiting 54.23 to 70.18 and 57.19 to 71.48 per cent reduction in pod and flower damage respectively. More over, this native *B.t.* isolates have shown more then 65% overall control of *Helicoverpa armigera* on cabbage and tomato crops and effectively protected their fruits from larval damages. However, they remained second best effective to the insecticide treatments. It can be concluded that *B. thuringiensis* can be an obvious choice due to their considerable efficacy as an alternative to chemicals or for their possible use in IPM against many insect pests, which will in turn became a viable proposition in mitigating the chemical residues on agricultural crops as well as for breakdown in building up of pesticide resistance in insect pest.

Future Thrusts

- More native *B.t.* isolates are to be found out for specific group of insects, which are fully compatible to our agro-ecosystem.

- Strain improvement and genetic modification of native isolates are to be carried out for a broad-spectrum biopesticide, which can give desired control of polyphagous insects and other pests of agricultural crops.
- Presently, in Indian market *B.t.* transgenic cotton is available, which should be replaced with native transgenic of regionally preferred cotton varities. Moreover, this technology will be more useful for vegetable and pulse crops, which have only 1 or 2 serious pests. *B.t.* transgenic for such crops is urgently required to minimize health and environment hazards.

—•—

Chapter 18

Entomopathogenic Nematodes (EPNs) A Novel Tool for Management of Insects, Phyto-Nematodes and Diseases of Agricultural Crops

INTRODUCTION

Entomopathogenic nematodes (EPNs) especially members of genus *Steinernema* (27 species) and *Heterorhabditis* (8 valid species) are innovative bioagents for plant protection scientists of India. These EPNs have symbiotic bacteria (Genus *Steinernema–Xenorhabdus* and Genus *Heterorhabditis–Photorhabdus),* which are gram negative, facultative anaerobic rods belonging to enterobacteriaceae having dimorphism nature. These mutually associated bacteria cause mortality of host insect within 24-48 hours of nematode entry in to the host and having wide host range amongst class Insecta. They are also found safe to non-target organisms and compatible with pesticides. Symbiont also produces antifungal and antibacterial metabolites like Xenorhabdin, Xenocoumacins, Xenoxodes, Nematophines (3′ indole ethyl 3′ methyl -2′ Oxo) and soluble proteineous compounds which make EPN a broad spectrum bioagent for integrated pest and disease management (Vyas, 2000).

Management of insects through EPNs during last two decades is increased to a surprising level in developed countries and now EPN based several biopesticides are widely marketed in Europe and America as well as considered as second most adopted biopesticide after *Bacillus thuringiensis* (Table 18.1). An effort on use of EPN in India was started since late 60's but till day it is under extensive research phase. Few workers in India have tested EPNs proving them very effective and useful for insect management during last decade. Natural incidence of *Steinernema* and *Heterorhabditis* is also recorded from Simla, Gujarat, Nilgiri hills, Coimbatore, Bangalore and Delhi. Ecomax Pvt. Ltd., Mumbai had brought two internationally marketed products of Bio Sis Company for marketing in India viz. Soil commando (*Steinernema* sp.) and Green commando (*Heterorhabditis* sp.). These products were also tested through multi location trials in the country during 1995-97. Recently in last few years research efforts through AICRP on Biological control, PDBC, Bangalore, is initiated country wide on testing of EPNs in different crops for management of key insect pests. Efforts carried out on EPN research at Gujarat Agricultural University (Now, Anand Agril. Univ.), Anand are also highlighted in details this review.

INDIAN SCENARIO OF EPN RESEARCH UP TO 2000 AD

Narayana and Gopal Krishnan (1987) reported better efficacy of DD 136 at Banglore against larval and pupal stage of *S. litura* at the dose of 100 IJS and above that percent mortality in laboratory conditions.

Das and Divaker (1987) have studied compatibility of DD136 with pesticides at Hyderabad and found few key agrochemicals compatible with EPN. Rao *et al.* (1975) tested the toxicity of various Organophosphate insecticide to IJs in water suspension and obtain 90 -100 % mortality after 24 h. Ghode *et al.* (1988) reported effective management of *H. armigera* with use of *S. feltiae* at different doses, 10-10000 IJS / ml with 29 to 70.63 % control under laboratory condition at Bhubaneshwar. Bijjur *et al.* (1991) studied compatibility of *S. feltiae* and *H. armigera* NPV at Dharwad.

Shivkumar *et al.* (1992) reported 50 - 29.24% laboratory control of *S.litura* by use of *H. bacteriophora* at the dose of 20 IJS from Coimbatore. Shanti and Shivkumar (1992) studied Effect of EPN on Red hairy caterpillar with good field control on groundnut crop.

Vyas and Yadav (1993) tested efficacy of EPN @ 1000 IJS/m^2 against *H.consanguinae* and reported 40-100 % infection at 10 & 20 after DAT on groundnut crop at Anand. Vyas and Yadav (1993) observed efficacy of *S.glaseri* @ 16 IJS/ml and 32 IJS/ ml against *S.litura* and *A. ipsilon* and reported 100% compatibility at Anand. There was a 90-10% infection of Steinernema spp was observed in citus butterfly *P. demoleus* on citrus plant at IIHR,Bangalore (Singh, 1993). From Ludhiana *S.feliae* (N-25) reported to infect *Pieris brassicae* Oryzacphillies at the dose of 500-2000 IJS/5 days by Mathur *et.al.* (1994).

Patel and Vyas (1995) reported that the *S.glaseri* was found effective at the dose of 4000 Ijs/ 20 plant on different host on chickpea crop and reported 29.7% infection at Anand. Shivkumar *et al.*(1996) tested efficacy of *S. carpocapsae* at the dose of 500 IJS / flower head against *S.litura* and reported 21.7% reduction in insects at Coimbatore.

The efficacy of *S.glaseri* was tested against *P.xylostella* at the dose of 100,000 IJS/m^2 on cress at Anand and reported 64.1% Higher yield. (Annon.,1997) Shivkumar *et.al.* (1996) studied efficacy of *S.carpocapsae* at the dose of 500 Ijs flower head in sunflower at coimbatore aginst *S.litura* and reported 21.7% reduction in insect population.

H. indica was found effective against *Leucinodes orbonalis* and reported 4.65 to 12.2 infection of host and reduced damage up to 47 percent. *S. carpacapsae* observed 8.45 to 20.2 percent infection in host and help in reducing fruit damage.

Joseph Rajkumar and Shivakum (1997) studied efficacy of *Steinernema sp.* against *S.litura* at Madurai on Blackgram and reported 1500 IJS / 96 h plant. Mathur *et al.* (1997) found that compatability of *H. bacteriophora* with carbaryl in laboratory against *S. litura* at Delhi.

Patel (1999) carried out the laboratory bioassay of two *Steinernema* spp. against *H. armigera* and found that after 120 hrs *Steinernema* sp. (VI) induced maximum mortality (79.96%) at 50 IJs/larva were applied. Whereas, significantly low larval mortality (49.96%) was observed in 10 IJS /larva treatment and no mortality in control. Similarly, in *Steinernema* sp. (AI) bioassay the *H. armigera* mortality was upto 49.96% at 50 IJs dose after 24 hrs, which was followed by lower doses. But at 120 hrs maximum 87.99% larval mortality was recorded at highest EPN doses (40 and 50 IJs /larva).

Ganguly and Singh (2000) reported *S. thermophilum* has broad host range and *S. thermophilum* found highly pathogenic at New Delhi. Hussaini *et al.* (2001) also observed cent percent larval mortality of *A. ipsilon* and *A. segetum* by *indica* in filter paper assay.

EPN RESEARCH AT ANAND AGRICULTURAL UNIVERSITY

Research work was initiated in 1990 with an imported EPN, *S. glaseri* (Steiner) Travassos (SBI, Coimbatore) by testing it's pathogenicity against few polyphagous insect-pests viz., white grub (*H. consanguinea* Blan.), cut worms (*Agrotis ipsilon* Huf. and *S. litura* Fab.) and gram pod borer (*H. armigera* Hub.) through laboratory, pot and field trials on bioefficacy of *S. glaseri* at AICRP on Biological control of crop pests and weeds and subsequently in much detail at Department of Nematology.

Laboratory bioassay of *S. glaseri* against *A. ipsilon* and *S. litura* produced cent percent larval mortality at 16 and 32 IJs/g soil doses after 72 h exposure respectively (Vyas and Yadav, 1992). Laboratory evaluation against *H. consanguinea* indicated that grub mortality had significant correlation to nematode doses over 3 days period ($Y = 14.4 \times 1.05X$, $P < 0.01$), whereas in small field plot condition, nematode application at different doses gave 40 to 100 % suppression of grubs on groundnut (Vyas and Yadav, 1993). But. *S. glaseri* was able to induce only 29.7 % mortality of *H. armigera* larvae on sixth day of treatment at highest dose tested (4000 IJs/20 plants) on chickpea in a pot study (Patel and Vyas, 1995).

Up to 1994 our research efforts on insect management through EPN were not reached to the field conditions due to lack of suitable *in vitro* production technology. But in two research projects carried out at Department of Nematology, GAU, Anand (1994 - 2000), ample research work on isolation of native EPNs (Vyas *et al.*, 1998), *in vitro* mass production technique (Vyas *et al.*, 1999) and field testing of EPNs was carried out. Through Post Graduate Research also, EPN work was further strengthened in GAU (Patel,1999; Patel and Vyas,2001). The salient findings of both the projects are summarized here after.

NATIVE MASS PRODUCTION TECHNOLOGY FOR ENTOM - PATHOGENIC NEMATODES, *STEINERNEMA / HETERORHADITIS*

In vivo multiplication of EPN

Mass production of EPNs isolated from soil was carried out by charging 3rd instar larvae of *Corcyra cephalonica* under laboratory conditions (plate II). Five host larvae were released in a plastic tube lined with filter paper on bottom of the tube and 1 ml nematode suspension having 100 IJs was inoculated on the filter paper

(20 IJS/larva). Mortality of larvae was recorded at 24 h interval up to 7 days. For emergence of EPN from dead larva cadavers were kept on moist white trap. Emerging IJs were collected daily and stored at 25± 2°C in BOD incubator.

In vitro mass production of EPN on Solid Media

Infective juveniles (IJs) of EPNs emerged from dead host larva were collected in 0.1 % formalin prepared in D/W and surface sterilized by suspending in 70 % methyl alcohol prepared in D/w for 15 minutes. IJs were then rinsed with sterile D/W for 2 times.

Media composition

- 100 g of pulverized dry dog food
- 500 ml dissolved 1% agar

  OR

Egg Yolk Yeast Agar medium

- Hen Egg Yolk 10%
- Yeast Extract 1%
- Agar Agar 2% (Vyas *et al*, 2001)

Media were prepared in 100 ml distilled water with Agar - agar as solidifying agent and were sterilized at 121°C for 10 min. in autoclave and poured in 5 cm diameter glass Petri plates. A well of 7 mm diameter was prepared in the centre of Petri dishes having solidified media using a sterile cork borer under aseptic conditions. *In vivo,* mass produced fresh IJs were inoculated @1000 IJs/plate in the wells using sterile pipette in laminar flow chamber (Table 18.2, Vyas *et al.,* 2001). Plates were kept individually in plastic container (8 cm D) having D/W in surroundings to collect emerging IJs. Plates were than incubated at 27± 1°C in BOD incubator. After 6-7 days harvest freshly emerged IJs and stored at 27± 1°C in BOD.

IN VITRO MASS PRODUCTION OF EPN ON LIQUID MEDIA IN POLYURETHANE SPONGE

Isolation and maintenance of symbiont bacteria

The entomopathogenic nematodes, S. *glaseri* and one unidentified *Steinernema* sp. were cultured in the laboratory on host insect larvae,

C. cephalonica, following white trap method (Poinar, 1979). Infective juveniles (IJs) of EPN, emerging from dead host larva, were collected in 0.1 % formalin solution in distilled water. These were then surface sterilized by suspending in 70% methyl alcohol (prepared in sterile distilled water) for 15 min followed by rinsing (twice) with sterile distilled water. The treated IJs heamolymph was streaked on Mac-Conkeys agar plate under aseptic conditions for isolation of symbiont bacteria, *Xenorhabdus* spp. The plates were incubated at 27 ±1 °C for 24- 48 h for development of colonies. Isolated pure bacterial cultures were further streaked on a differential medium, NBT agar (Akhurst, 1986). Gram staining as well as gelatin liquefaction test for assessment of purity of the culture was carried out. The cultures were preserved in deep freeze at O°C and regular subcultured at 30 days internal.

Production of Infective Juveniles on Artificial Media

The symbiont, *Xenorhabdus* bacteria were first mass cultured in 0.5% yeast sodium chloride broth for 24 h with vigorous shaking of the tube 3-4 times. Bacteria (10^5) were then inoculated on 1.5 cm^3 size polyurethane sponges (6 sponges/flask) having artificial medium (a modified medium of Bedding, 1984) in 250 ml flasks, previously sterilized at 10 psi for 10 min in autoclave.

Medium Composition

Chicken extract (10%) -10 ml	(Freshly prepared)
Yeast extract 0.5 g	(Himedia, India)
Beef extract 0.5 g	(Hi media, India)
Distilled water 9.0 ml	
pH -6.8 -7.0	

(Vyas *et al.,* 1999)

Flasks were shaken and kept at 27 ± 1°C in BOD incubator for mass multiplication of *Xenorhabdus* in sponges, for 24 h. After 21 hr surface sterilized 5000 IJs were introduced in the flask and after 15 days interval, nematodes from sponges were harvested using 500 mesh (25μ) sieve. Harvested IJS were preserved in 0.1% formalin in a refrigerator at 4-10°C with aeration on alternate days.

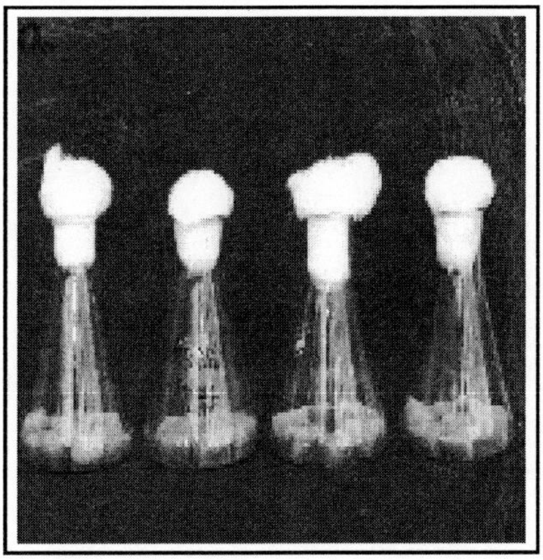

Multiplication in polyurethane sponges with modified chicken extract liquid media.

↓

Multiplication on egg yolk solid media and Granular formulation.

Xenorhabdus nematophilus subsp.

↓

Granular ofrmulation

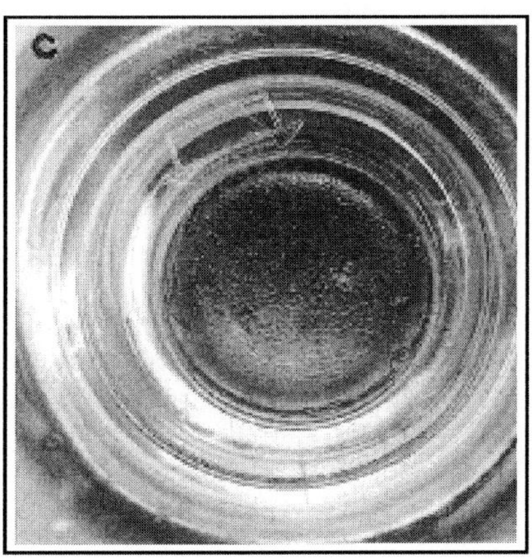

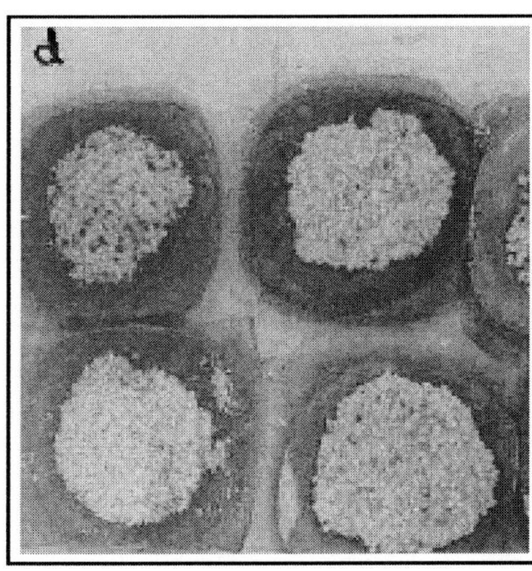

Fig 18.1. *In vitro* multiplication techniques

Research Achievements Nematodes (*Steinernema*) in AAU

Two years pooled results on bioefficacy of *S. glaseri* against white grubs in pots revealed that laboratory augmented *S. glaseri* infective juveniles (IJs) @ 2000, 2500 and 3000/pot effectively protected groundnut plants giving 41.5 to 68.7 % mortality of white grubs.

Testing of *S. glaseri* against *A. ipsilon* in pots revealed maximum 66.0 % mortality of larvae induced by nematode application @ 3000 IJs/pot after 8 days of treatment (Patel *et al.*, 2001).

Two years pooled results on bioefficacy of EPN against *S. litura* indicated that coating of antidesiccant agents like gum arabic (GA) alone @ 5 % and its combination (50:50) with liquid paraffin (LP) and/or starch (ST) on *S. glaseri* IJs proved most effective in inducing quick and higher mortality of larvae compared to untreated (IJs) (Vyas *et al.*, 1997).

An *in vitro* mass production technique of *S. glaseri* on symbiont bacteria, *Xenorhabdus nematophilus* using liquid media having fresh chicken extract supplemented with beef and meet extracts on polyurethane sponges has been standardized since 1994 for quick and easy reproduction of IJs for laboratory and field trials (Vyas *et al.*, 1999).

Two years pooled results on biological suppression of white grubs using EPN on groundnut in micro plots indicated that maximum mortality of *H. consanguinea* was recorded in phorate treatment (91.25 %) that was at par with IJs either with starch or liquid paraffin antidesicant treatment followed by *S. glaseri* alone treatment (71.25%).

Three years pooled results on bio-efficacy of EPN against Diamond Back Moth, *Plutella xylostella* L. on cress indicated that EPN alone and coated with antidesicant (LP+GA) enhanced larval mortality by 14.77 and 32.20 % respectively over initial population at 7 days of treatment, succeeding treatment was at par with treated checks, *B. thuringiensis* (76.14 %) and monocrotophos (53.06%) treatments. Both the EPN treatments also increased cress seed yield by 43.53 and 125.13 % respectively over control. However maximum yield was recorded in *B.t.k.* treated check.

Demonstration trial on bioefficacy of *S. glaseri* with and without antidesicant against DBM *P. xylostella* on cress showed good reduction of larval population with 20.3 and 29.7% increase in cress yield over control, however *B.t.k.* and monocrotophos treatments remained more effective than EPN treatment. ***(Scientific community recommendation of Gujarat Agricultural University, 35th Plant Protection Sub committee Meeting, held on April 3-4, 2000).***

Management of *Helicoverpa armigera* using entomopathogenic nematodes

Sixteen entomopathogenic nematodes were isolated from Central and North Gujarat for management of *H. armigera* (11 *Steinernema* spp. and 4 *Heterorhabditis* spp.). Amongst the isolates, EPN-3 (Anand) and EPN-16 (Dehgam) were found, highly virulent to *H. armigera* larvae, while remaining isolates induced low mortality of the larvae under laboratory bioassay (Vyas *et al.*,1998).

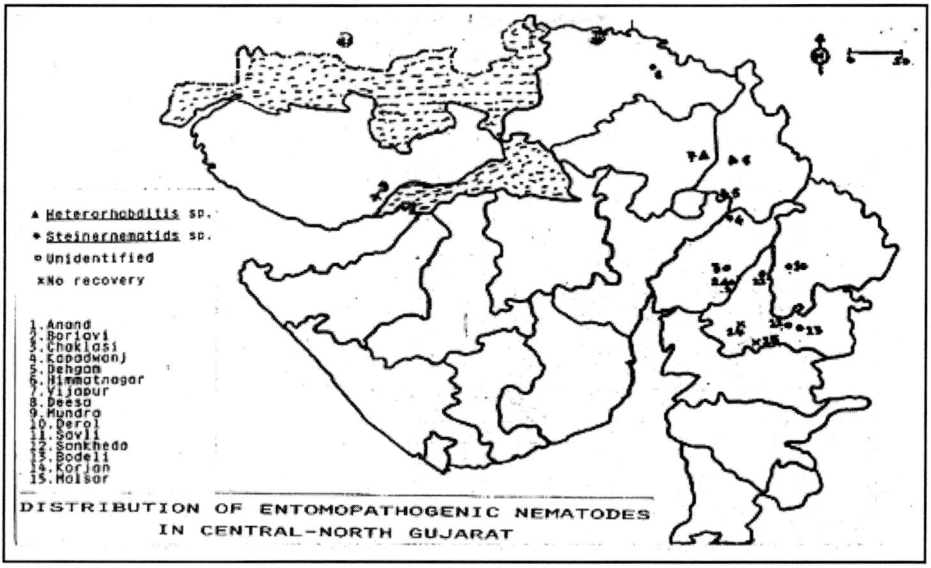

Fig 18.2. Distribution of entomopathogenic nematodes in Gujarat

Bioefficacy of promising isolates EPN-3 and EPN-16 carried out against *H. armigera* on pigeonpea and chickpea induced maximum 96.8 and 70.9 % larval mortality respectively @ 2000 IJs / 5 plants / pot treatment (Vyas *et al.*, 2001b).

Amongst several artificial media tested, animal protein originated (10) and plant protein originated (8), Hen egg yellow medium was found best for reproduction of *Steinernema in vitro* yielding (Av. 12,13,816 IJs/ 5 cm D petriplate). The cost of production for 1 million *Steinernema* IJs (material only) comes to 72 paise in popular media, dog food biscuit agar, where as cost of production in modified egg yolk agar media comes to 25 paise only (Vyas *et al.*, 2001a) (Table 18.2).

Application of EPN alone and with adjuvant @ 1 lack IJs/m2 on cotton Hy.6 reduced bollworm damage to fruiting bodies and increased the seed cotton yield by 20.18 and 14.09 % respectively over untreated check (Vyas *et al.*, 2001c).

Application of EPN alone and with adjuvant significantly reduced *H. armigera* on pigeonpea compared to initial population. Both the treatments reduced 31.7 and 16.8 % pod damage with 66.9 and 97.3 % increase in yield over control (Vyas *et al.*, 2002b).

Both the native EPN isolates, EPN-3 and 16 were found effective for management of *H. armigera* on chickpea (Vyas *et al.*, 2002a).

Sodium alginate based EPN formulation preparation was successfully carried out in the laboratory but longevity of IJs was observed to be affected on drying and hardening of prepared alginate encapsulated material.

EPN-3, GAU - Anand and Mogar isolates are characterized as *S. riobrave* Cabanillas, Poiner & Raulston, 1994. (Anonymous, 2001; Ganguly *et al.*, 2002)

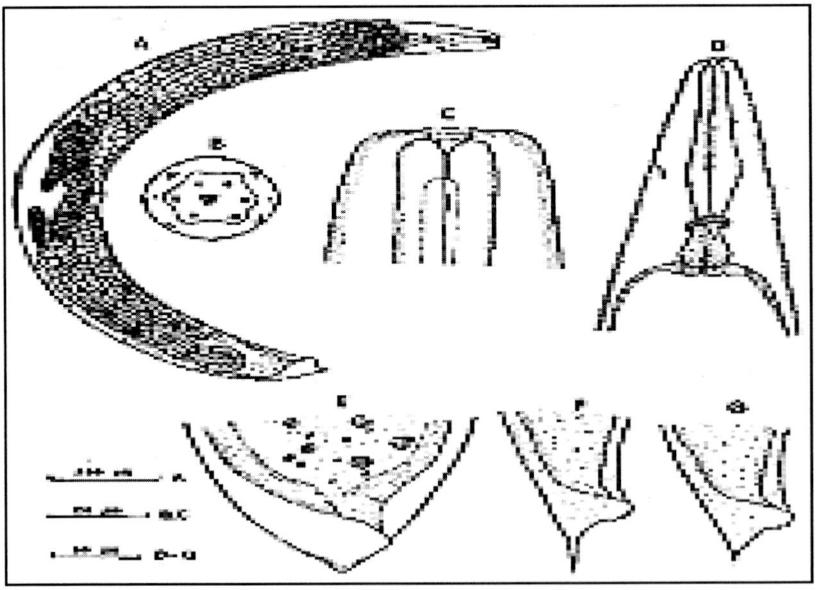

Fig 18.3. *Steinernerma riobrave*

Potential of native EPN, *Steinernema* sp. and symbiont bacterium, *Xenorhabdus* sp. for management of pest complex through biotechnological approaches.

In vitro effect of *S. riobrave* (GAU EPN M & A isolates) against RKNs (*Mj* pt. 2 & *Mi*) in cavity blocks was carried out in the laboratory respectively. Results of both the experiments showed that EPN has no direct impact against RKNs like predatory nematodes. But higher mortality of RKNs in bioassay during initial hours was suspected due to release of secondary metabolites of symbiont bacteria of *S. riobrave* from dead IJs (Maghodia *et al.*,2003).

Two years pooled data revealed that sequential and simultaneous application of EPN and RKNs (*M. incognita*) on okra cv GOH 1 in pots showed that RKI / plant was low in EPN - RKN (one week) treatment followed by EPN + RKN simultaneous treatment indicating that *S. riobrave - Xenorhabdus* complex has suppressive effects on RKNs (Table 18.3, Vyas *et al.*,2003).

Interaction of *S. riobrave* and *M. javanica* pt. 2 on groundnut showed that average number of galls (RKI) / plant was significantly low in EPN - RKN treatment followed by simultaneous application of RKN + EPN and RKN - EPN treatments. Probably the *S. riobrave* - *Xenorhabdus* sp. complex has suppressive effects on *M. javanica* pt. 2.

This suppose to be the first indicative report on suppression of a plant parasitic nematodes by beneficial entomopathogenic nematode in India (Vyas *et al.*, 2004). This is in Confirmation to EPN Vs PPN Suppression phenomenon (Table 18.3, Fig. 18.4).

Laboratory bioassay of *S. riobrave* against white grub, *H. consanguinea* indicated that after 48 hrs, there was no effect of *S. riobrave* against *H. consanguinea* larvae, however the increased exposure time invariably increased the larval mortality. After 168 hrs, the maximum *H. consanguinea* mortality (44 %) was recorded at 40 IJs / g soil which remained at par with 20 IJs / g soil.

Two years pooled results on efficacy of *S. riobrave* (different inoculum levels) against groundnut pest complex (*M. javanica* pt2 and *H. consanguinea*) in pots indicated that low RKI was found in carbofuran treatment followed by EPN treatments both being statically at par compared to control. Significantly low white grub was recorded in carbofuran and EPN treatment @ 1,00,000 IJs / m^2 (592 IJs/pot). ***As a result S. riobrave gave satisfactory control of white grub and M. javanica pest complex on groundnut in pots inciting a broad-spectrum biological control tool*** (Vyas *et al.*, 2005b).

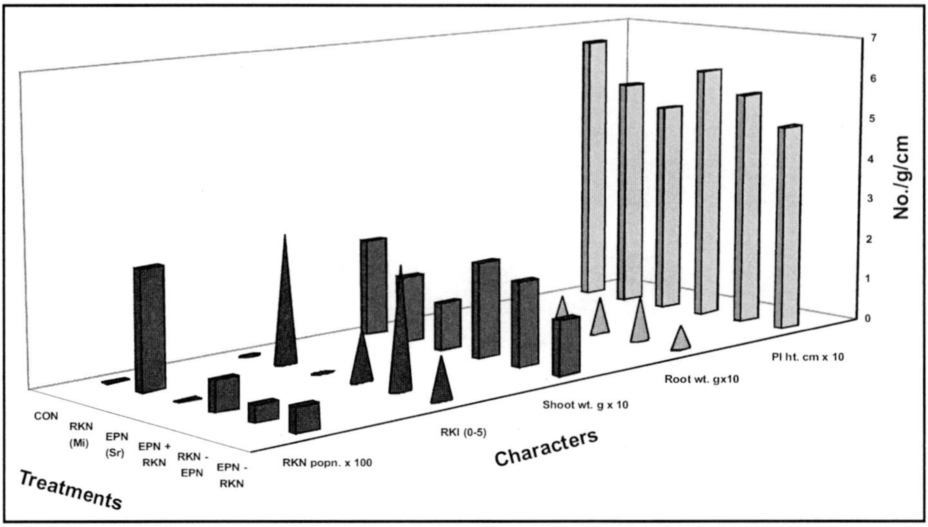

Fig 18.4. Interaction of EPN and RKN on GROUNDNUAT in Pots

All native *Xenorhabdus* isolates matched with only few biochemical and fermentation characteristics of Family *Enterobacteriaceae* in general, but have more similar reactions of Genus *Xenorhabdus*.

Both exo and endo toxins of *Xenorhabdus* spp. showed some suppressive effect against *M. javanica* pt. 2 on okra and against *M. incognita* and *M. javanica* pt. 2 on tomato in pots *(Studies on target pest on alternative crop hosts for confirmation of phenomenon)*.

The average no. of galls / plant (RKI) was significantly low (1.52) in *Sc Xenorhabdus* and *St Xenorhabdus* exotoxins applications followed by *Sr* M *Xenorhabdus* exotoxin in pot bioefficacy against *Mj* pt. 2 on groundnut. Similarly, effect of endo toxins of different isolates against *M. javanica* pt. 2 revealed that root - knot index (RKI) was minimum in OH *Xenorhabdus* (1.38) followed by *Sr* M *Xenorhabdus* endo toxin (1.52), which remained at par with *Sc Xenorhabdus* (1.63) and *St Xenorhabdus* (1.72) endo toxins.

In vitro toxicity of *Xenorhabdus* metabolites following poisoned food technique of all the isolates against *A. niger* showed toxic effect (fungistatic) in undiluted form, but 1:100 dilution was tolerated by the fungus.

First confirmative report on use of entomopathogenic bacterial metabolites for control of plant pathogenic fungus in India (Fig. 18.5. Vyas *et al.*,2005a).

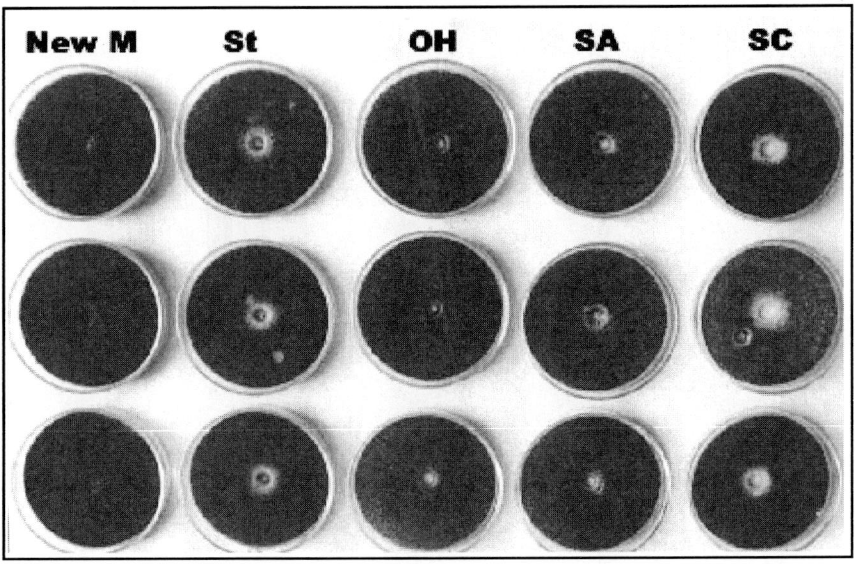

Fig. 18.5.

Studies on *Xenorhabdus* metabolites of EPNs against collar rot disease of groundnut revealed that seed coating with exotoxin proved effective in suppressing collar rot fungus (*Aspergillus niger*) on groundnut. The endo toxins of all isolates exhibited equal efficacy against *A. niger*. Among EPN isolates, *Sc* endotoxin exhibited higher plant height followed by OH and *Sr* M isolates.

Soil application of *Xenorhabdus* metabolites in root zone of groundnut plants in pots indicated that soil inoculation of exotoxin proved effective in suppressing collar rot fungus on groundnut. Similarly, the endotoxin of all isolates exhibited equal efficacy against *A. niger* (Table 18.4).

The results on total exo-protein analysis showed that all the four natively isolated *Xenorhabdus* spp. of different *Steinernema* spp. produced wide variety of different secretory proteins mainly in three clusters having molecular weight in the range of 20 - 21 Kilo Daltons (KD), 46 - 51 KD and 60 - 66 KD. The proteins of high molecular weight, 76 - 90 KD apart from regular proteins that are not produced in *Sc* and *St* might have some additional benefits (Vyas *et al.*, 2006b).

The genomic DNA (total) bands seen on agarose gel indicated similarity between native *Xenorhabdus* isolates and Ohio standard culture.

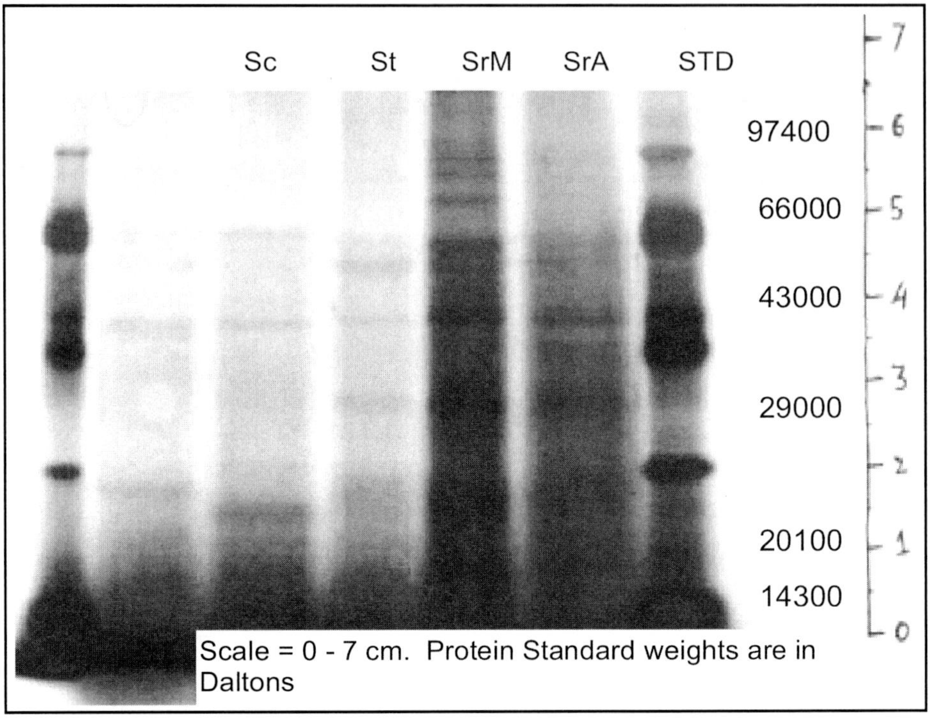

Fig. 18.6.

PCR amplification of genomic DNA showed multiple amplified products suggesting a possible new subspecies of *Xenorhabdus* (*Sr* A & M, *St*) from the Indian subcontinent (Vyas *et al.*, 2006b).

Pot efficacy of native *Steinernema* spp. (*Sr, St* & *Sc*) against groundnut pest complex showed that the lowest root-knot index (RKI) was observed in carbofuran (1.49) treatment which was statistically at par with EPN *Sr* M (1.58) @ 1 lakh IJs/pot and *P. lilacinus* fungal bioagents (1.51) treatments. Significantly low white grub population was recorded in EPN *Sr* M treatment, which also remained at par with *M. anisopliae* and carbofuran treatments (Vyas *et al.*, 2006a).

Field bio-efficacy of native *Steinernema* spp. against groundnut pest complex revealed that EPN treatment, *S. riobrave* (Mogar) isolate had minimum RKI (1.97) and was at par with *M. anisopliae* (1.90) and other EPN isolates. Significantly higher green pod and fodder yields were also recorded in *Sr* M treatment as compared to other EPN isolates. However, carbofuran being nematicide remained significantly superior over all the bioagents.

Table 18.1. International Market of EPNS.

EPN	Trade name	Company	Target	Note
H. bacteriophora	**Cruiser**	Ecogen. Inc USA	Soil dwelling coleopteran & Lepidopteron	Registration exempt
	Nematop, Nema green	GmbH, USA	Black vine weevil, grubs	Not necessary
H. Megidis	Dick	Andermalt Biocontrol- AG Swiss, Austria	Black vine weevil	Registered
	Larva nema	Koppert Biological system B.V., USA	Vine weevil	Canada/UK/USA/SA
Steinernematid	Biosafe	SDS Biotech K.K., Japan	Coleopteran/ Lepidopteron	Registered Japan, SA
Steinernema & *Heterorhabditis* spp.	Scanmask, Ecomask, Heteromask	Biologic Company, USA	Soil & Cryptic insects	No registration for US
S. carpocapsae	Bio vector, Vector TL, Savior, Biosafe	Thermo Trilogy Corporation, USA	Cutworm, Armyworm, Root weevil, Vine weevil, Stem borers	US, EPA exemption
S. feltiae	Nemaplus	Andermalt Biocontrol AG Swiss, Austria	Sciarid flies of mushroom Sciarid flies	Registered Canada, USA, SA
	Magnet X-Gnat	Thermo Trilogy Corporation, USA	Mushroom flies	USA, EPN exemption
S. riobrave	Biovector 355, vector MC, Devour	Thermo Trilogy Corporation, USA, Ecogen Inc. USA	Citrus root weevil complex, Mole crickets, Plant parasitic Nematodes	USA, EPN exemption

Source: faded on http://edis.ifas.ufl.edu/IN096, www.nysaes.cornell.edu

Table 18.2. *In vitro* production cost of EPN (Native technology)

Media	Material cost (Rs.)			Media plates prepared	Ave. production/ plate	Total production	Cost for 1 million IJs production (Rs.)
	Egg Yolk(15g from one egg)/ Biscuit (2 g)	Agar agar powder cost	Total cost				
Eggyolk 10% +agar 3&%	1.80	7.5	9.80	30 plates (150ml)	1.2 million	36 million	0.25
Dog food biscuit 2% + agar 3%	0.50	5.0	5.50	20 plates (100ml	0.38 million	7.6 million	0.72

Note: 1. All the equipments, glasswares, plastic wares etc. are common for IJs production the cost is only worked out based on media used for comparison purpose.

2. 1 million IJs is sufficient to treat 10 m2 area of crop/soil for foliage/soil insect pests.

3. Composition of Hen egg : Av. wt. 50-52 g, Yolk (Yellow) 31%, Albumin (white) 58 % and shell 11 %. From one egg 15 g yolk is collected which is used to prepare 150 ml media

Table 18.3. EPN vs PPN

EPN	PPN	Crop and Reduction	References
Steinernema carpocapsae	*M. incognita*	Tomato Significant reduction	Kermarrec *et al.*, 1991
Heterorhabditis bacteriophora	*M. incognita*	Tomato reproduction factor decreased	Kermarrec *et al.*, 1991
Steinernema & Heterorhabditis (Six species)	*M. incognita* *Pratylenchus coffeae*	Suppressive on root invation by J2 (76 to 97 %)	Matsanga *et al.*, 1997

Adopted from Anonymous 2005

Possible Mechanism of Entomo-pathogenic nematodes (EPNs) against Plant parasitic nematodes (PPNs)

- Competition of space : *Sterinernematides* have affinity towards plant roots and over crowing in rhizosphere may push out PPNs.
- EPNs symbiotic bacterium have natural biostatic compounds act as repellents to PPNs (Further studies required).
- Predator Pray Response : Soil predators consumes PPNs & EPNs andapplication of EPNs is soil increases predators (Highly reproductivly 5 days/generation) Evidences are gathered to support this phenomenon.

Over all fact is EPNs are marketed and advocated as for management of Plant Parasitic Nematodes in several countries and EPNs are reducing PPNs populations in pots, turf grass and fields is real (see table - 18. for some initiative reports from Japan).

Table 18.4. Antibiosis of *Xenorhabdus* and *Photorhabdus* metabolites on plant parasitic fungi. *(Few selective reports from different part of world)*

EPN symbiotic bacterium	Toxic compounds/ metabolites	Plant Pathogenic Fungi	Reference
Xenorhabdus nematophills	Nematophin (3'indole ethyl 3' methyl-2' Oxo)	*Phytopthora boehemeria* *P. infestans*	Li-jan Xiong *et al*, 1997
X. bovienii	Soluble metabolites	*Pythium,Rhizoctonia* Spraying@5mg/ml reduced blight by 24 % on potato	Yong xiu Fen *et al.*, 1998

Contd...

Table 18.4. Contd...

X. nematophillus (Bacteria phase I)	Antimycotic substances	Antagonistic to 32 species of fungi like *Pythium, Phytophora, Mucor, Botrytis* etc	Chen *et a.,l* 1994
Photorbabdus luminescens	Antimycotic substances	*Pythium, Phytophora, Mucor, Botrytis*	Chen *et al.,* 1995
X. bovienii (Phase I & II)	5 x 10^6 / ml	Damping off fungus of sugar beet *Pleospora bectac.*	Chen *et al.,* 2000

Adopted from Anonymous 2005

Interaction of EPN, *S. riobrave* and RKN (mix population) on okra cv. GOH 1 in nursery beds indicated that EPN has a suppressive effects on RKNs on okra roots. Lowest RKI / plant was recorded at 1 lakh IJs / m^2 which remained at par with carbofuran @ 1 kg/ha followed by fungal bioagent, *P. lilacinus* @ 25 kg/ha. Significantly, higher okra yield was obtained in carbofuran treatment followed by *P. lilacinus* and EPN 1 lakh IJs / m^2 treatments followed by lower doses of EPN.

Efficacy of EPN, *S. riobrave* and their toxins against RKN (mix population) on okra cv. GOH 1 indicated that EPN had a suppressive effect on RKNs on okra roots. Lowest RKI (1.44) / plant was recorded in *Sr* M *Xenorhabdus* isolate which was at par with fungal bioagents (1.62) followed by carbofuran (1.80) treatments. Significantly higher okra yield was recorded in carbofuran treatment, which remained statistically at par with fungal bioagents followed by EPN *Sr* M, and other EPN isolates (Vyas *et al.,* 2005c).

Future Thrusts

- Isolation of more and more native EPNs from different states of the country with search for virulent strains / species for different groups of insects which are having adaptation to local climates and also ecologically well compatible.
- Development of easy and cheap mass production / fermentation technique with suitable formulations for commercialization of indigenous isolates. Low cost industrial mass production technique with greater virulence of augmented EPN.
- Training of scientists (AICRP/State level) for handling, testing, maintenance and mass multiplication of EPNs. Research efforts through networking programme are also needed.

- Genetic improvement of native isolates for greater virulence and desiccation tolerance.
- Biotechnological studies to be strengthened on symbiotic bacteria and their genetics for development of transgenic plants at a long run.

—•—

Chapter 19

Nuclear Polehedrosis Virus (NPV)

Baculoviruses are a group of viruses that kill particular insects including crop pests. The value of different baculoviruses as environmentally safe biopesticides for forests and pests of agricultural crops has been well established o ver last five decades. In the late 1960's the **nuclear polyhedrosis virus** (NPV) was studied through series of tests in India and thorough out the world as new biopesticide belonging to insect pathogenic baculoviruses. The registered pioneer viral insecticide was containing this NPV produced for lepidopteran caterpillars. A number of other NPVs have been partially tested. But their commercial competitiveness with chemical insecticides lead to swifter death or cessation of host insect feeding. It is also perceived that the practical use of baculoviruses depends on a satisfactory outcome to thorough assessment of the release. . In addition, the rapidly expanding discipline of Invertebrate Pathology has failed to find incidence of NPVs and GVs infecting hosts outside the few host ranges. The NPVs proved harmless to non-insect invertebrate cell lines vertebrate cell lines, vertebrates, plants and non-arthropod invertebrates. In the field, baculoviruses can persist in the soil for at least 3 years. Birds are major pathway by which baculoviruses are spreads; 40% of individuals and 9 bird species found to excrete active virus on feeding with infected insects. Between 1 and 10% of baculoviruses remained on plants but the soil contained the remaining 90 to 99%. No baculoviruses were found in percolation water. Air flow, rain splash, parasites and aphids all dispersed viruses. Despite baculoviruses having many alternative lepidopteran hosts, field infections were not found in alternative hosts, suggesting no risk of environment pollution and other hazards.

Biology and replication of NPV

NPV life cycle involve two structurally distinct phenotype. Budded virus phenotype produce during early stage of infection and spread infection from cell to cell within insect and is highly infection from cultural cells. The polyhedra derived virions (PDV) phenotype (also known as occluded virus (OV)) is produced during terminal stage of viral infection becomes concentrated in the nucleus and is occluded within polyhedra. PDV are relatively non-initiate infectious for cultural cells. After ingestion by a susceptible insect polyhedral are dissolved releasing PDV which initiate infection within the midgut cells. In both early and late viral gene expression of glycoprotein is synthesized and transported to margin of the cell where it incorporated in to the host cell plasma membrane. During budding process the nucleocapsids become envelope by the modified membranes. PDV become surrounded by an envelop that assembles within the nucleus subsequently they become occluded and released upon the death of cell.

NPV Production

The lab reared or field collected larvae are reared on semi synthetic diet consisting of Chickpea grain flour, Antibiotics, Vitamins, yeast, preservatives in Multi cell cavity trays or plastic vials filled with semi synthetic diet. After solidification the diet surface inoculate POB's with syringe @ 20 μl/cell.3rd to 4th instar larvae are release in tray @ 1 larvae / cell and fix lid. Collect diseased larvae before liquification stage occurs to harvest entire amount of POB's formed. In case of field collected larvae the trays are observed for dead larvae after two days of inoculation and such dead larvae are removed. After five days watch the tray again and collect the dead larvae in container with distilled water in a ratio of 1 larvae : 1 ml. Remaining larvae transferred to other trays with fresh diet. The collected dead larvae are macerated at the end of one week in a warring blender using distilled water @ 2 ml per diseased larvae. Later the product put in to a centrifuge at 500 rpm for 2 minutes and settled debris is discarded. Products again centrifuge at 5000 rpm and residue settled is collected in fresh distilled water.

Standardize for POB counts with the help of haemocytometer under microscope.

$$\text{Number of POB} = \frac{D \times X}{N \times K}$$

D = Dilution factor, X = Number of POB count,

N = Number of squares,

K = Volume of one square in cm3. By counting number of POB's per larvae, larval equivalent can be determined.

Success Story of NPV Applications for Insect Control

Bioassay of NPV

The NPV of *Heliothis* (Ha NPV) is now developed as a commercial viral pesticide through various research and development phases world over since late 60's. In India, laboratory studies have indicated that NPV is most potent and virulent insect pathogen giving high mortality of test insect, which can be further increased by addition of certain adjuvant too.

Patel *et al.* (1968) reported NPV infection in gram pod borer for the first time in India and found that *H. armigera* larvae when infected with NPV became sluggish, stopped feeding, paralyzed and the integument became soft and brown to black with internal tissues disintegration and liquification in the laboratory.

Ethiraju *et al.* (1988) carried out laboratory evaluation of few NPV formulations against *H. armigera* larvae at Tamil Nadu. Talc based wettable powder formulation and unformulated NPV were equally effective against second instar larvae of *H. armigera*. But dust formulation was less effective.

Rabindra and Jayaraj (1988) evaluated certain adjuvents with NPV against *H. armigera* on chickpea leaves as food and they reported either addition of 0.5% Jaggery, 1% cotton seed, kernel flour, 3% groundnut oil cake and 1% chickpea flour significantly increased NPV induced larval mortality. The combination of various adjuvants like 0.25% jaggery, 0.25% chickpea flour, 0.5% groundnut oil cake and 0.1% cotton seed oil cake together have greatly enhanced the efficacy of the virus.

Rabindra *et al.* (1994) tested the influences of host plants (e.g. Chickpea, Pigeonpea, Cotton etc.) surface and the virulence of NPV against *H. armigera* larvae. The median lethal concentrations (LC_{50}) and median lethal time (LT_{50}) in 2nd instar *H. armigera* indicated that cotton leaf was most detrimental followed by chickpea and pigeonpea while lablab and sunflower leaves had no impact on effectiveness of NPV.

The effect of feeding of *S. litura* larvae for 24 hrs of castor leaves dibbed in suspension of combination of NPV and sub lethal concentration of DDT, Gama BHC, melathion and pyrethrin and bactospeine was determine in laboratory. Supplemental effect was observed when virus was combined with DDT at 5 ppm or pyrethrin at 10, 20 or 50 ppm. The combination of virus and bacterium produce only a sub-additive effect, whereas malathion found antagonistic to virus (Komolpith and Ramakrishna, 1978).

Mallathi and Shriramulu (2000) conducted laboratory experiment to test the bioefficacy of *Btk viz.*, Dipel, Delfin, Biofet, Biolep and Bioasp, at 0.075 %; The botanical extracts and Nemagold extract of *Azadiracta indica* at 0.03 %, the nematode formulation Green commandoes (*Steinernema* sp.) at 25 sponges/acre and endosulfan 0.07 % against 2nd instar larvae of *S. litura* and DBM (*P.xylostella*). The *Bt* formulations were equally effective against *P. xylostella* and *S. litura* resulting in 86.66 % to 100 % and 96.66 % to 100 % mortality respectively in 72 hrs.

Field use of NPV

Narayanan (1979,1980) has reported the effectiveness of NPV against *H. armigera* on chickpea from Karnataka, who applied NPV at 250 and 125 LE/ha during evening hours thrice at weekly intervals after the appearance of pests and recorded significant reduction in the caterpillar population. Jayaraj (1981) has also reported effectiveness of NPV against *H. armigera* on chickpea from Tamil Nadu, virus spraying at 125 LE/ha with endosulfan at half the recommended dose (0.035%) found effective and significantly reduced the pest damage to pods and grains.

Santharam *et al.* (1981) gave two applications of the NPV at 375 LE/ha and found maximum protection against *H. armigera* and also harvested high yield of chickpea. Field test conducted at Gujarat Agricultural University, Anand revealed that NPV when sprayed at 100 LE/acre in 350-400 liters of spray fluid for five to six times induced high mortality of *H. armigera* larvae ranging between 81.20 to 98.0% with an average of 89.80% in chickpea (Anonymous, 1983).

Rabindra *et al.* (1989) tested the efficacy of various adjuvants with nuclear polyhedrosis virus (NPV) to control *H. armigera* in chickpea. They reported that addition of crude sugar to NPV gave the highest yield (796 kg/ha) followed by whole egg homogenate and cotton seed oil being equally effective and at par to each other.

Pawar *et al.* (1990) compared the NPV and 0.05% endosulfan for management of *H. armigera* in chickpea and found that all the treatments significantly increased grain yield over check except the lowest concentration of virus alone.

Mishra *et al.* (1991) have conducted field trials for *H. armigera* suppression on chickpea in Bihar and Uttar Pradesh. They demonstrated that a single spray of 250 LE of Ha NPV in 500 liters of water gave 97.2% mortality of the pest.

Rabindra *et al.* (1992) reported that the mortality of *H. armigera* larvae due to NPV application was significantly higher on *H. armigera* susceptible varieties of chickpea than on resistant accessions (ICC 506 and ICC 10817).

Vyas and Lokhohoura (1996) evaluated HaNPV for the control of *H. armigera* on chickpea at Pantnagar. Three spray of NPV @ 250 LE/ha significantly reduced the larval population, however, two application of 0.07% endosulfan was found superior to NPV spraying.

Shenhmar *et al.* (1996) conducted an experiment to study the comparative bioefficacy of different dosage of *Ha*NPV and *Bt* formulations against *H. armigera*. Mortality of *H. armigera* due to *Ha*NPV was very low upto 72 hrs of treatment, compared to *Bt* formulations namely Delfin and Dipel.

Healthy

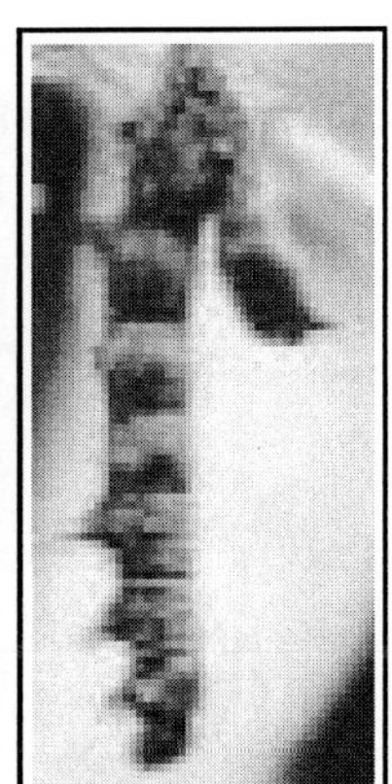

Diseased

Helicoverpa armigera on chickpea

Appearance

Insects killed by baculoviruses have a characteristic shiny-oily appearance as shown in above picture healthy insect left and diseased right. Often seen hanging upside down limply from vegetation. They are extremely fragile to the touch, rupturing to release fluid filled with infective virus particles. This tendency to remain attached to foliage and then rupture is an important aspect of the virus life-cycle.

Field persistency of biopesticide NPV

In an agro-ecosystems soil, moisture, temperature and sun light (UV radiation) are the three most important factors that influences the effectiveness of biopesticides. The persistence, accumulation and denaturation of microbial agents in the environment are very critical factors in determining the success of biopesticide. Much of the research undertaken during the past 2 decades in this area has been devoted for assessing the effects of various environmental factors are reviewed.

Carlo *et al.* (1976) studied persistence of HaNPV in laboratory and field. They found increased persistence Ca. 2 to 5 folds, under both simulated and natural sunlight.

Narayanan (1979) studied on leaching of Ha NPV from black soil. He reported that labeled polyhedra applied to the top of the soil column were detected by radiological means up to 10.2 cm in 18 days after continuous leaching.

UV protecting and spreading additives commonly used for NPV products

Talk, Kaolin, clay as diluent (dust)

Triton X - 100 as Wetting agent/spreader

Whole milk, Jaggery as phagostimulant

Robin blue, Tinopul, Congo red, Charcoal, Black ink as sunlight protectant.

Emerging Trends

Genetically improved NPV for greater virulence and persistence in field.

Toxic gene of *B.t.*, scorpion venom gene, mite neurotoxin gene and chitinage gene have been genetically engineered to NPV for making them more efficient.

The endotoxin gene has been cloned to produce transgenic plants with resistance to pests.

Widening the use of NPV genome as vector and cloning vehicle in genetic engineering.

—•—

Chapter 20

Entomopathogenic Muscardine Fungi Beauveria bassiana and Metarrhizium anisopilae

More then 100 fungi are reported world over belonging to various classes like, Ascomycota - sexually produced ascospores form within a saclike ascus (tiny spore containing sac); Basidiomycota - sexually produced basidiospores form on a supportive structure called a basidium; Deuteromycota - sexual cycle of reproduction unknown attacking insects. They had been recorded from 200 insect hosts belonging to orthoptera, hemiptera, lepidoptera, diptera, coleoptera etc. Among this muscardine fungi are having broad spectrum, more effective and widely used in last few decades and regarded as one of the most promising species amongst known entomopathogenic fungus imperfecti (Deuteromycota) for the development into a practical and potential insect biocontrol agent.

White Muscardine Fungus (*Beauveria bassiana*, Deuteromycotina: Hyphomycetes)

The white muscardine fungus, *Beauveria bassiana* (Balsamo) Vuillemin (Deuteromycetes) was observed in about 900 A.D on silkworms in Japan. More importantly A. Bassi in 1834 in Russia postulated the germ theory of disease by studying this fungus. *Beauveria* has a worldwide distribution and has a wider host range than the other Deuteromycetes infecting insects from most of the insect orders. Fungus is effective in controlling both bollworms and caterpillars as well as sucking pests. It has provided excellent control of many of the major insects of fibre, cereals, vegetables and ornamental crops.

Mode of Action (Mycoses)

Fungal conidia germinate on the insect cuticle by producing germ tubes, which penetrate the rough body wall. The infection propagule in muscardine fungi are the coniodiophore, which primarily invades in insect through the host cuticle, by the body wall, spiracles, and also through mouthparts. The penetration is both mechanical and enzymatic through the action of protinase, lipase and chitinase on the body wall. Growth of the fungus in the haemocoel is by budding, which produces hyphal bodies. These are transported throughout the haemocoel and give rise to localized concentration of mycelia followed by host death, due to depletion of nutrients and impact of fungal toxins (Ferron, 1978). A heavy growth of interwinding mycelia develops in the body cavity after 1-2 days of infection (Fig. 20.1 below) and dead insect becomes completely mummified. Finally hyphae emerges from the dead cadaver, sporulation occurs on the host surface which imparts white OR green coloration with cushiony growth all over dead insect (see plate 2 & 3).

Fig. 20.1 shows that infection starts with conidia getting in contact with insect cuticle, they germinate and produce penetration structures, germ-tube, appressorium, physical obstruction or invasion of organs and toxicosis (Destruxins etc.).

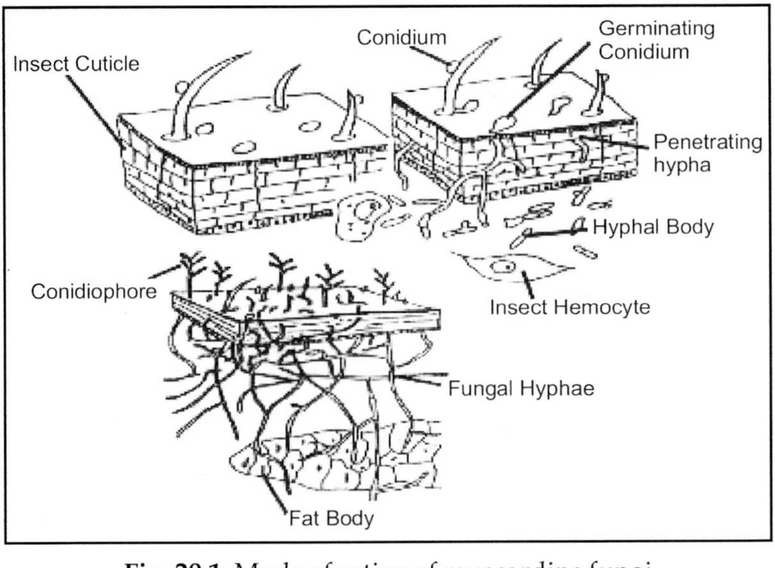

Fig. 20.1. Mode of action of muscardine fungi

Efficacy of White Muscardine Fungus

Bio-assay of white muscardine fungus, *B. bassiana* against *H. armigera* and other related lepidopterans pests have been successfully cited by many scientists. Ignoffo *et al.* (1979) reported that three species of caterpillar, *Trichoplusia ni* Hub, *Plutella xylostella* L. and *Pieris rapae* L. of cole crops were susceptible to boverin, a mycoinsecticidal preparation of *B. bassiana* exhibiting LC_{50} values, 0.25, 0.025 and 0.27% respectively. Prasad *et al.* (1989) tested the susceptibility of tobacco caterpillar, *S. litura* Fab. to three entomogenous fungi and indicated that *B. bassiana* (Bapatla isolates) was most virulent giving lowest LC_{50} as 19.9 x 10^5 conidia/ml. Moreover they found that susceptibility of larvae were decreased with increase in larval age from 2nd to 4th instar in respect of LC_{50} and LT_{50} values. Gopalakrishnan and Narayanan (1990) studied the dose - mortality relationship between *B. bassiana* and *H. armigera*. They reported that fungus was pathogenic to all the stages of *H. armigera* inflicting 60 to 100% mortality in I -V instar respectively as well as 100% mortality of eggs at various doses tested. Whereas, pupal mortality was 60% at 1 x 10^9 conidia/ml and 80% mortality in pre pupal and adults stages at 1.0 x 10^{10} conidia/ml, with an incubation period ranging from 2 to 14 days.

The white muscardine fungus, *B. bassiana* (Bal.) Vuill. has been reported from India attacking several pests like cabbage semilooper, rice stem borer and leaf eating caterpillers, brinjal ash weevil, banana leaf beetle and gram pod borer etc. Rao (1975) conducted a survey on microbial diseases of rice pests at Rice Research institute, Cuttak and found *B. bassiana* infection in rice stem borer and leaf eating caterpillars. Ignoffo *et al.* (1979) studied boverin; a mycoinsecticidal preparation of *B. bassiana*. Boverin (5.0 %) application equivalent to 7.4 x 10^{14} conidia/ha effectively reduced field population of *T. ni*. Gowda and Raviprasad (1992) reported the occurrence of *B. bassiana* on the redgram pod borer, *H. armigera* and found that relative humidity 86.6% favoured maximum fungal infection, while at 58.3% RH it was lowest. Moreover in the study highest mean maximum temperature (29.8 °C) induced 11.12% infection in larvae, whereas at lowest mean maximum temperature (25.4 °C) little higher infection 12.93% was noticed. Saxena *et al.* (1997) tested *B. bassiana* for the control of *H. armigera* infesting chickpea and found very effective results. Spraying of 2.6 x 10^7 spores/ml resulted in only 6% pod damage with 2377 kg/ha yield, while untreated control recorded very high as 16.3% pod damage with 1844 kg/ha yield. Vandenberg *et al.* (1998) tested the *B. bassiana* spores for control of diamond back moth on crucifers and found that treatment and insect stages (2nd, 3rd and 4th instar) both have significant effect on survival of larvae. Ramesh *et al.* (1999) proved the efficacy of *B. bassiana* @ 2.7 x 10^7 against *Syllepte derogata* on sorghum and reported effectiveness of fungus against the pest.

Green muscardine fungus : (*Metarrhizium anisopliae* (Metschnikoff) Sorokin, Deuteromycotina: Hyphomycetes)

Metschnikoff (1879) described the fungus from wheat grain beetle, *Anisoplia austriaca*. Then the species of M. anisopliae & *M. flavoviridae* has been identified. The fungi causes disease by adhesion & formation of germ tube, produces toxins & enzymes. The green muscardine fungus *M. anisopliae* has been reported from India attacking several pests like borers, loopers, semiloopers, some sucking pests etc.

Efficacy of Green Muscardine Fungus

Gopalakrishnan and Narayan (1988) in laboratory bioassay found that entomofungal pathogen *M. anisopliae* (Metschnikoff) Sorokin var. *minor* Tulloch was very effective against *H. armigera.* Jun *et al.* (2000) bio-assayed isolate FI1248 of *M. anisopliae* against larvae of *P. xylostella.* Larvae were exposed to cabbage leaf discs after being immersed in to conidial suspension. Nine days after inoculation, the LC_{50} was between 4.17 x 10^5 - 6.33 x 10^5 conidia/ml. With a dose of 10^7 conidia/ml concentration, the LT_{50} was 4.7 days. The results showed that isolate FI1248 was highly virulent to *P. xylostella*. Pharindera (2000) tested *B. bassiana* for *H. armigera* and recorded its effectiveness from the fifth day of treatment giving highest 31.67 per cent larval mortality. Laboratory and field tests were conducted to determine the potentials of *M. anisopliae* for the control of cabbage pests. Two strains, Fm8 and Fm12 showed high virulence to the larvae of *P. xylostella* under field and laboratory conditions both. However, strain Fm8 was judged as slightly more potent than strain Fm12 based on larval mortality and yield (Santiago., 1991). Shelton *et al.* (1998) recorded good efficacy and persistency of *M. anisopliae* and other fungi against DBM on cabbage seedlings. Fan *et al.* (1999) conducted a field survey on the entomopathogenic diseases infecting *H. armigera* in 3 localities in Hebai and China during 1995-97. The results showed that the main pathogenic organisms were HaNPV, HaCPV, HaGV, *Bt, M. anisopliae* and *B. brongniartii*. The rate of control of *H. armigera* by those organisms varied greatly with year and generation of the insects. The average mortality of *H. armigera* was 21.2 per cent and could reach 57.1 per cent in an epidemic year.

White and Green Muscardine Disease of White Grub and Root-Knot Nematodes

Pathogenicity studies of *B. brongniartii* against white grubs in laboratory revealed that grubs of *H. serrata, H. consanguinea, H. fregei* and *A. nathani* were highly susceptible to fungus. The fungus also attacked the adult,egg and pupal stages. Efficacy of fungus tested against *H. consanguinea* and *H. serrata* grubs in groundnut field @ 10^{15} conidia/ha resulted in 41.5 and 45.5% disease in the grubs respectively. Efficacy of *M. anisopliae* against *H. consanguinea* - various stages of different species of white grubs available were tested against the fungus in the laboratory. Almost all species were found susceptible to muscardine infection. A small scale trial was conducted in groundnut crop to determine efficacy of *M. anisopliae* against *H. cosanguinea* indicated more than fifty per cent suppression of grubs in field condition (Vyas *et al.*,1990). This has suggested a new direction and very good opportunity for cost effective, long term white grub management using insect pathogens as an alternative to chemical insecticides in India. Moreover, the muscardine fungi, *M. anisopliae* and *B. brongniartii* (potential bioagents for the control of soil insects) were reported to parasitized *M. incognita* eggs (Vyas *et al.*,1995). These two fungi were also tested individually and in combination with *P. lilacinus* against *M. javanica* pt.2 and white grubs attacking groundnut in pots. Muscardine fungi were found to manage white grubs (upto 85 %) and RKNs as well and these fungi were also found effective against *M. javanica* pt.2 on groundnut under field condition (Vyas *et al.*,1997).

Interaction of Fungal Biopesticides

A desirable characteristic of microbial agents is the ability to attain high amount of disease in the target insect population. To achieve this, numbers of laboratory experiments have been conducted world over in which simultaneous or sequential release of two or more pathogens are advocated.

The compatibility of *Beauveria* with five commercial preparations each of insecticide fungicide and botanical pesticide commonly used in crop protection was established to utilize the bioagent in Integrated management of white grub. Amongst the insecticide tested Phorate was found compatible with fugnus, whereas Lindane, Quinalphos, Carbofuran and Sevidol showed moderate to severe inhibition of the fungus. In case of fungicides the Fytolan , Captan and Mancozeb at 100 ppm concentration did not impair the growth of fungus, whereas Carbendazim inhibited fungal growth completely. Thiram showed moderate inhibition (Vyas *et al.*,1990). Amongst the botanical pesticides the Neemark (Azadiractin) was found compatible to fungus, while Nicotin Sulfate , Replin , Indiara and Phytoallexin showed moderate inhibition of the fungus(Vyas *et al.*,1992). Sporulation of the fungus was adversely affected by fungicides but not by the insecticides The fungus did not show any antagonistic effect on *B. popilliae* and exhibited mix infection in grub under laboratory conditions.

Mass Culturing and Maintenance of Muscardine Fungi

M. anisopliae, conidia ellipsoidal to rod shape, strong hydrophobic. Both the fungus is able to produce blasto- spores and conidio -spores on verity of substrates. *B.bassiana*, conidia ellipsoidal to round, hyaline, medium hydrophobic.

Stock culture should be maintained on selective media (Dodine and Chloremphenicol supplemented, Veen and Ferron,1986) used for the routine maintenance of isolate and for production of conidia potato dextrose agar is to be used. For mass production PDA plates (50) are inoculated under sterile conditions @ 10^4 spores/plate from stock fungal culture. But before that dodine (10 μg/ml) is added by the time of pouring of media in to plates increased the efficacy of recovery of muscardine fungi and inhibits common fungi. PDA dishes are incubated at 25-27°C under B.O.D. incubator. To reduce the dehydration of the media, petri dishes were sealed with parafilm. Within 8-12 days, the culture is get sporulated and the conidia were harvested in sterile D/W 5ml/plate by scrapping from the surface using a sterile glass spreader.

Fungal blasto-spores obtained in submerged culture are much less stable than conidia and are consequently difficult to process after harvesting. Laboratory cultures are frequently freeze-dried with or

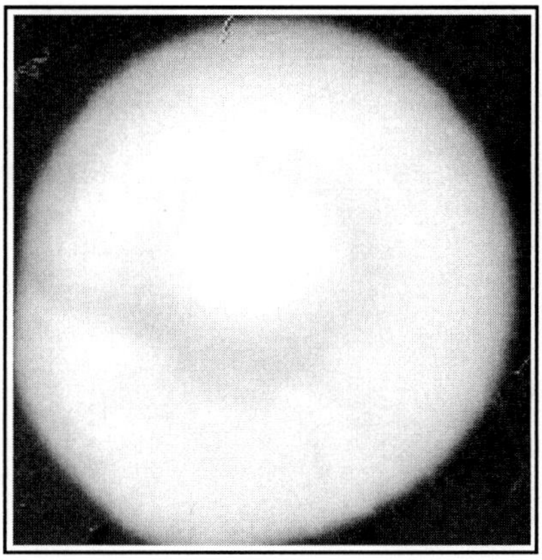

Plate 20.1. *B. bassiana* colony on agar plate (close up).

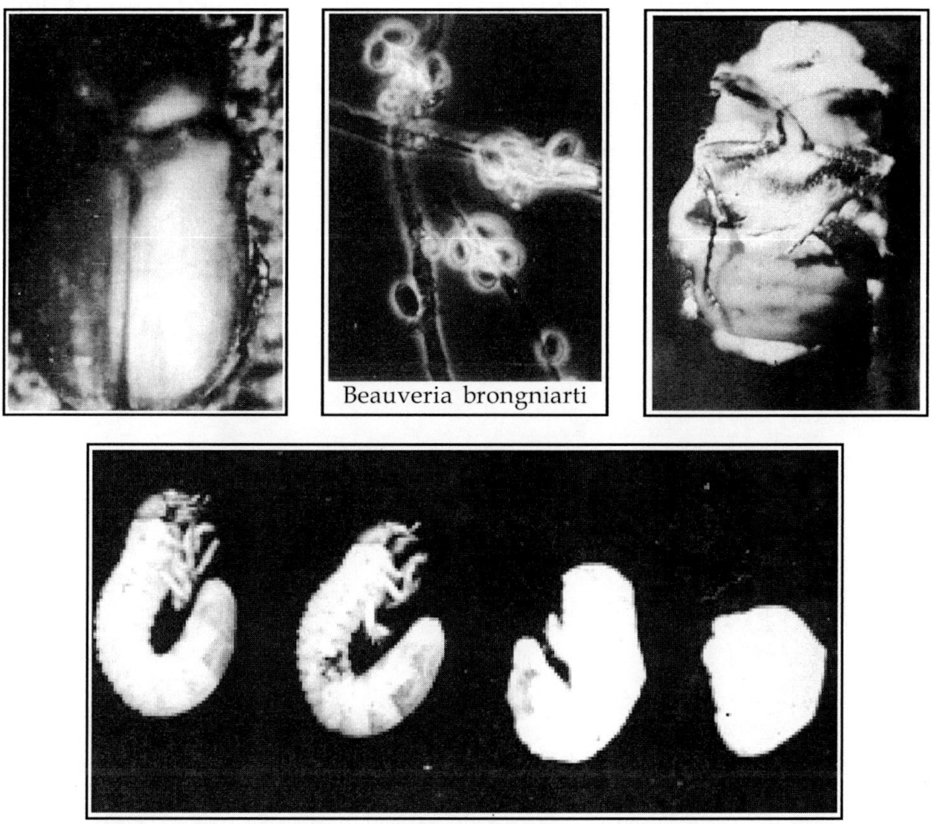

Plate 20.2. White muscardine fungus on white grub (root grub) adult beetle and larvae, healthy and infected respectively with fungal microphotograph showing conidia.

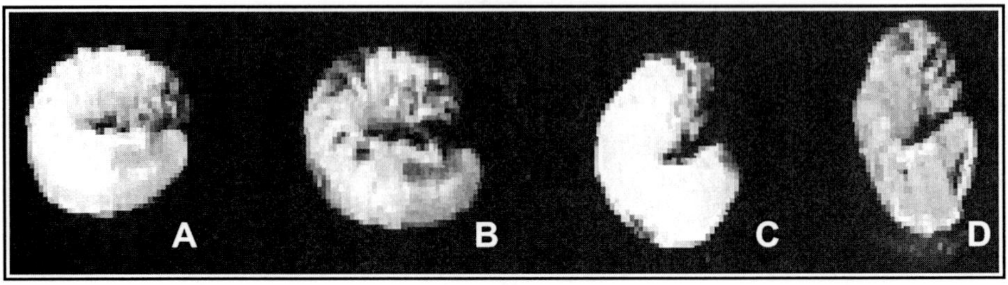

Plate 20.3. Green muscardine fungus on white grub (root grub) healthy (A) and infected larvae (B-D).

without protectants, but even dried product may have short viability hence, majority of production attempted for conidio-spore production. *B. bassiana* product in five different ways: vacuum, freeze, spray-drying, drying by mixing with inert filler, and in a fluidized bed with an inlet temperature of 40°C and an outlet of 30°C. The virulence of the fluidized bed-dried material is enhanced and the process accelerated by precipitation using a calcium carbonate, surfactant, silica gel mixture. In addition, a sulphite liquor is mixed in. Drying in a vacuum desiccator produced samples of high viability and virulence, some remaining active for 1 year at 4°C storage. Spray-dried *B. bassiana* spores together with the culture media led to a complete loss of viability. However, using a 2 % molasses mixture as a protective medium permitted retention of viability and efficacy, although to a lesser extent than for the freeze-dried product. The spray-dried conidia showed no loss of viability up to 18 months storage at cool and low temperatures.

The suitability of various cereal grains and agricultural wastes made it very for mass production of muscardine fungus. Crushed maize grains/rice found an ideal substrate for maximum sporulation of the fungi (Vyas *et al.*,1991). High density polyethylene autoclavable bags (HDPE) were found very suitable and cheap for small scale solid substrate fermentation of the fungus as described in details in for production of *P. lilacinus* in **Chapter _16_.**

Future Thrusts

The inoculative application of fungi alone or in integrated methods in India will certainly help in bringing down insect population below economic threshold level in due time, creating natural epizootics of the muscardine diseases. More toxilogical data on native isolates are to be generated for safety of muscardine fungi to mammal's, pet animals, beneficial insects, plants etc. are required for more and more adoption of bioagent for Public and field use.

—•—

Chapter 21

Microbiological Techniques

STAINING OF MICROORGANISMS

Various staining techniques are used to study morphology of microorganisms.

BACKGROUND

Staining is the method of artificially providing colour to microorganisms to allow the visualization under the microscope. Stains are employed not only to make the organisms visible but also understand their structure and chemical nature.

Dyes or stains are synthetic chemical products of the aniline type. They can be divided into acid dyes and basic dyes with respect to their colour property in the anionic and cationic form. An acid dye is the salt of a coloured acid while a basic dye is the salt of a coloured base. For example, Crystal violet, Methylene blue are basic dyes and Acid fuschsin and Erythrosin are acid dyes. Acid dyes have stronger tendency to combine with the cytoplasm whereas basic dyes have greater affinity for the nuclear region of the cells. Basic dyes are generally used in bacteriological studies.

Knowledge of the basis for the specificity of dyes is important to understand differential staining reactions. For example in Gram's staining, the primary dyes are of the basic triphenyl methane type (crystal violet or methyl violet). Alkaline aqueous solution of these dyes readily penetrates the bacterial cells or yeast and stain them dark blue.

If the stained cells are washed, extensively with water or dilute acid much of the primary dye is removed. Instead, if the cells are next treated with iodine solution which acts as mordant water insoluble dye complex is produced. When the cells are then treated with an organic solvent (alcohol or acetone) the dye iodine complex is quickly dissolved and tends to be removed from the cell contents whereas blue colour persists longer in the cell walls of Gram positive bacteria.

Application of an aqueous counter stain results in staining the decolourized Gram negative cells whereas Gram positive cell wall do not react with the counter stain. Evidently the essential difference between Gram positive and Gram negative microorganisms is in the chemistry of the cell wall, however, some of the environmental factors can also influence the staining property of the cells. They are (a) Age of culture (b) pH of the environment (medium). The response to Gram Stain is of value in classifying bacteria.

Gram Staining

This staining helps to differentiate between Gram +ve and Gram-ve cells, i.e. Rhizobia are Gram-ve.

Gram's Iodine

Dissolve 1 gm iodine and 2 gm Potassium Iodide in 300 ml distilled water.

Safranin Solution

Dissolve 0.25 gm Safranin in 10 ml 95% ethyl alcohol and mix with 100 ml distilled water.

Ammonium Oxalate Crystal Violet

Dissolve 2 gm crystal violet in 20 ml 95 % ethyl alcohol and mix with 80 ml 1% aqueous ammonium oxalate.

Staining Procedure

1. Prepare smears and fix in flame.
2. Stain with ammonium oxalate crystal violet for 1 minute
3. Wash in tap water.
4. Immerse in Gram's iodine for 1 minute
5. Wash in tap water and blot dry

6. Decolorize with 95% alcohol for 30 seconds.
7. Wash with water and blot dry carefully.
8. Counter stain with Safranin for 5 minutes.
9. Wash in tap water.
10. Dry and examine under microscope.

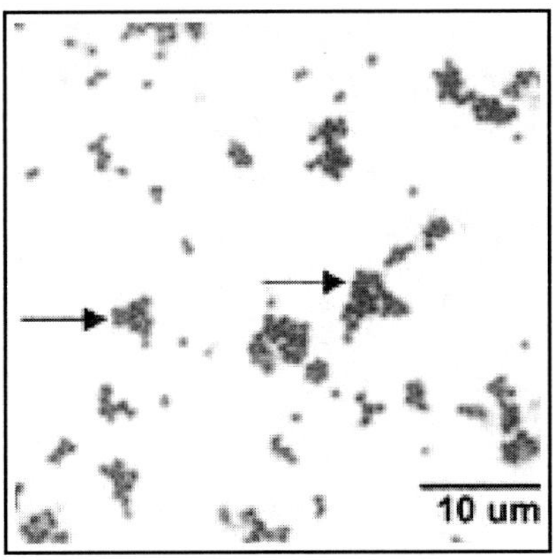

Gram + ve bacteria

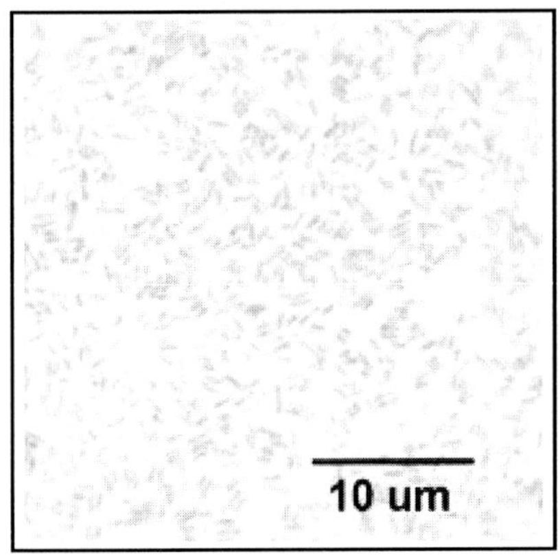

Gram - ve bacteria

Capsule Staining

The cell walls of many bacterial species are surrounded by a polymeric substance referred to as a capsule (if discrete) or slime later These capsules perform numerous physiological functions, including conferring resistance to phagocytosis and allowing adhesion to sites with favorable environments. The size of such capsules varies with species and among strains of a species, and may also be affected by the composition of the growth medium.

Because most capsule materials are water soluble, simple stains will not adhere to them, thus making capsule staining a difficult process. As a result of this limitation, capsule staining works by staining the bacteria and background more intensely than the capsule itself. Capsules are clearly visible in the light microscope only when they are negative stained or special stain are employed.

STAINING METHOD

Hiss's Method [Anthony's Modification]

Principle

Unlike the cell, the capsule is nonionic and primary stain can't adhere on application of basic dyes, cell pick up the stain; but when such stained cells are treated with decolorizing agent 20 % $CuSo_4$, it remove the excess primary stain as well as color from the capsule. At the same time, the copper sulphate acts as a counter stain by being absorbed into the capsule and turning it light blue.

In this procedure, smear should not be heat fixed since shrinkage is likely to occur and create a clear zone around the bacterium, which might be mistaken for a capsule.

Requirements

1. Capsulated culture of *Azotobacter.*
2. 1% Crystal violet stain and 20 % Copper sulphate solution.

Staining Procedure

1. Prepare a thin smear, don't fix it with heat.
2. Flood the smear with crystal violet stain for 4-7minutes.
3. Rinse the slide thoroughly with 20% $CuSO_4$ solution.
4. Allow the slide to air dry and examine.

   Result

   Cell appears dark violet while capsule are light violet in color.

MOTILITY OF BACTERRIAL CELL

Demonstration of the motility of bacterial cell.

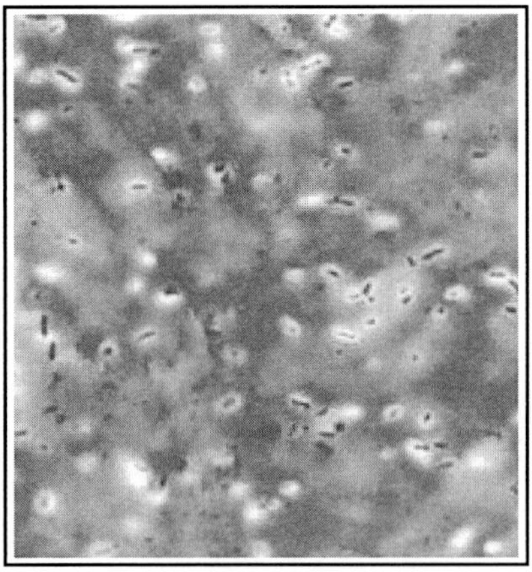

Capsule Surrounding cell

BACKGROUND

Heat fixing and staining are rather severe treatment for a bacterial cell and they cause considerable changes in morphology. To avoid these changes, bacteria can be observed in their living state by means of wet mount technique. In the method there is usually little difficulty with highly motile organisms but feebly motile organisms may require prolonged observation of individual cells. Bacterial movement must be distinguished from "Brownian movement" Brownian movement is vibratory movement caused by invisible molecules striking the bacteria displacing them for short distance. True motility of the bacteria may lead to the bacteria to move from on place to other relatively longer distance. Therefore, it is best to use hanging drop preparation for this purpose. For observing motility of pathogenic organism it is safer method to inoculate semi solid medium and observe growth and motility.

Materials Required

A broth culture of motile bacteria, *Azospirillum* sp., Microscope, Single cavity slides, Cover slips, Vaseline, soft agar medium.

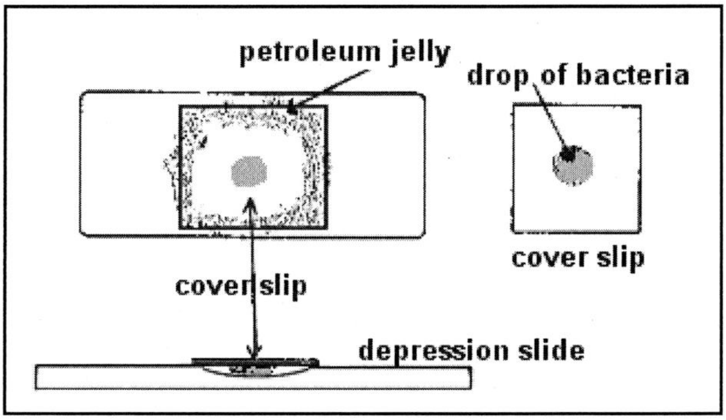

Procedure

Hanging Drop Method

1. Place a very small drop of bacterial culture in the center of a cover slip.
2. Place a small drop of Vaseline at each corner of the depression side of the cavity slide.
3. Invert the slide (cavity side) over the cover slip. The cover slip adhere to the glass slide and when the cover slip is inverted, the hanging drop is suspended in the well.

4. Bring the edge of the hanging drop into the focus with 10X objective before turning to the high power objective to observe motility.

PREPARING AGAR SLANTS

1. Prepare the agar medium in flask.
2. Keep it in hot plate (or) Oven at 80-90°C for ½ hour and then transferred (5 to 7 ml) to sterilize test tubes and plug with cotton.
3. Sterilize all the tubes at 15 lb for 20 minutes using a pressure cooker (or) autoclave.

Allow the test tubes containing the agar medium to cool down to about 45 to 50°C and keep them in a slanting position by resting the plugged end over the glass rod on the table and leave the tubes in this position until the medium has cooled to room temperature and solidified to look opaque. Store these tubes for later use.

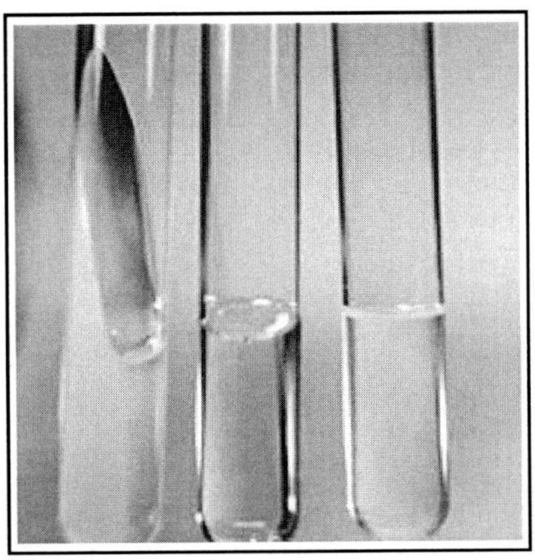

N. Agar slant, N.Broth, N.Agar Stab

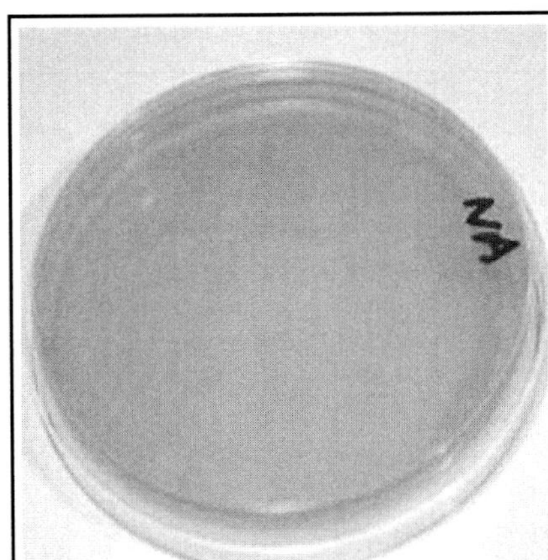

Nutrient agar Petri plate

PREPARING AGAR PLATES

1. Label the petriplates and organize them at inoculation table.
2. Grasp the Erlenmeyer flask containing sterile agar medium in liquid form with your right hand.
3. Turn your left hand palm side up and clamp the cotton plug between two fingers.
4. Remove the plug and flame the mouth of the Erlenmeyer flask.

5. Use the hand holding cotton plug to lift the lid of the petridish. Now pour about 20-25 ml of sterilized nutrient agar medium. Hold the lid so that it covers the bottom of the dish as you pour. This helps to prevent microbes and air borne dust particles from dropping into your sterile plate and contaminating it.
6. Immediately replace the lid.
7. Flame the mouth of the flask and pour the next plate in the same way.
8. If bubbles occur on the surface of the medium, break them aseptically by quickly passing the Bunsen flame over the surface.
9. Allow the plate to cool at room temperature (agar medium becomes opaque) and store the plates.
10. Observe all the liquid and solid media after two days. Properly sterilized medium will remain clean. The broth, which was not properly sterilized, would have developed turbidity. Use only media that are properly sterilized for your work.

PURE CULTURE OF BACTERIA

Pure culture technique and isolation of bacteria.

BACKGROUND

The microbial population in our environment – air, soil, water – is large and include many species of bacteria, fungi and algae. A study of the microorganisms in these habitats requires knowledge of the specific microbes present. This requires techniques for preparing species as pure cultures from the mixed population.

Pure culture represents a population of organisms of single species in the absence of living cells of any other species. There are various techniques whereby the different species in a natural population can be isolated and grown as pure culture. They are: (a) Streak plate technique, (b) Pour plate technique and (c) Spread plate technique.

Materials Required

Nutrient agar, Nutrient broth, Petriplates, Pipettes, Dilution blank, L-rod, tubes, Vortex mixer.

A. Streak Plate Technique

1. Use the nutrient agar plates prepared as earlier.
2. Sterilize the inoculation nichrome wire loop by flaming it red hot and allow it to cool for 30 seconds.
3. Holding the culture tube in your left hand take the tube near flame and remove cotton with your right hand and flame the mouth of the tube for a few seconds.
4. Touch the culture with the needle loop and take the culture.
5. Place it on the agar plate and streak it on the plate once.
6. Resterilize the loop and cool it as above and take the culture from one end of the plate to other end and complete the streaking.

Transferring the Inoculum into a Petri plate

1. Streak the inoculum over area "1." Flame the loop and cool it by sticking it into the edge of the agar.
2. Rotate the plate so area "1" is on your left. Streak the inoculum on area "1" over area "2." Flame the loop and cool it by sticking it into the edge of the agar.
3. Rotate the plate so area "2" is on your left. Spread the inoculum on area "2" over area "3."

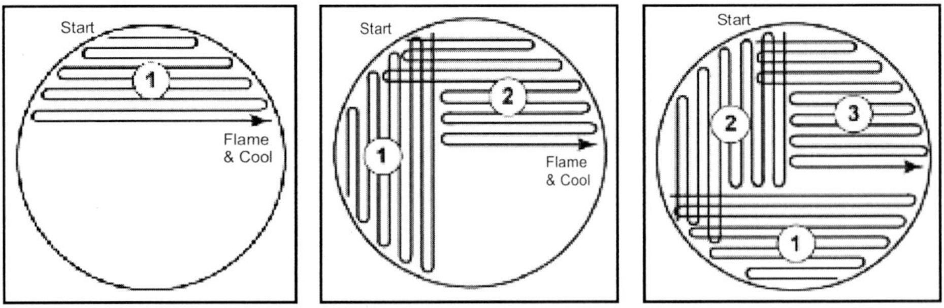

Keep the inoculating loop parallel with the agar and rotate the plate so you are always streaking the agar in the "12:00" position.

B. Pour Plate Technique

1. Liquefy the nutrient agar in the tubes by heating them in water bath.

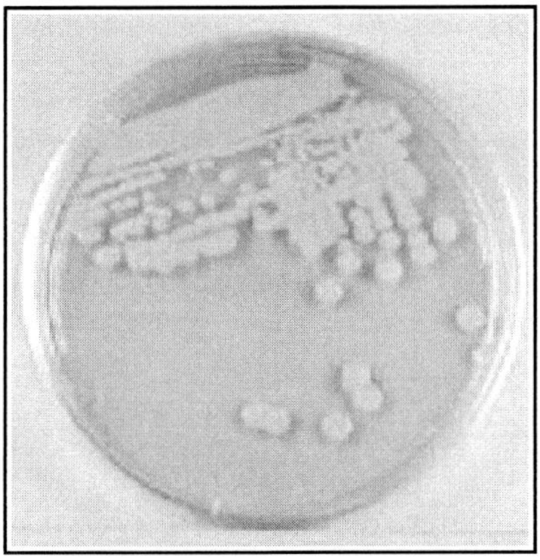

Plate Culture of *Bacillus* sp.

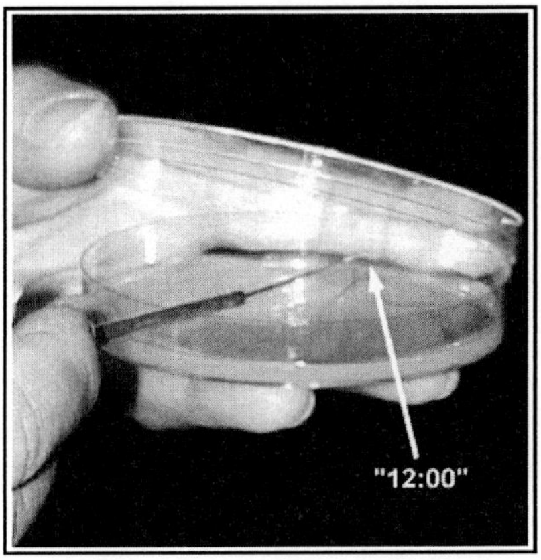

Streaking an Agar Plate

2. Cool the tubes to 45°C and hold at this temperature until ready to pour into the plate.
3. Label the tubes and corresponding Petridishes.
4. Serially dilute the given mixed culture and from each dilution take 1 ml of the sample.
5. Mix the sample with agar medium by gently rotating the tube between your palms.
6. Pour the contents of the tubes into the corresponding labeled Petridishes and allow to solidify.
7. Incubate the plates in inverted position at room temperature.

C. Spread Plate Technique

1. Use the nutrient agar plates as prepared in your earlier experiment.
2. Serially dilute the given culture as given in plate technique.
3. Label the nutrient agar plates corresponding to the dilutions.
4. Place 0.1 ml from the each dilution in separate nutrient agar plates.
5. Sterilize the 'L' rod with alcohol and then by flaming it.
6. Cool the rod and plate the rod gently on the surface of the agar, Spread the sample over the surface of the agar medium by

rotating the petriplate manually or by placing it on the rotating plate disc.

7. Incubate the plates in inverted position at the room temperature.

Observe the well-separated colonies from these different techniques. Pick up isolated colonies with the sterile inoculation needle and inoculate into the liquid cultures and observe them as before. After sufficient growth make wet mount of the cultures and observe under microscope. If the cultures are pure (axenic) they may be individually transferred to agar slants and labeled.

ASPTIC TRANSFER OF PURE CULTURES

Techniques for aseptic transfer of culture and inoculation.

BACKGROUND

A pure culture contains only one species of microorganisms. If another species of microorganism is accidentally introduced into a pure culture, the culture is said to be "contaminated" and is called a mixed culture. The appearance of contamination is always problem in microbiological technique because microorganisms are present everywhere. To avoid contaminations ensure (1) That all materials that will touch the culture are first Sterilized (2) All media to which the cells are transferred should be sterile, and (3) No contaminants are allowed to enter the growing culture. The methods of obtaining and maintaining pure cultures are collectively called aseptic technique.

Materials Required

Inoculation loop, Agar slant culture, Test tubes, Nutrient broth, Cotton wool, Microscope slides.

Following are the general principles of aseptic technique for starting transfer of cultures.

1. The growth medium with its container must be sterilized as soon as the medium is prepared.
2. The container must be plugged with cotton or covered with a cover to prevent entry of microorganisms.
3. Never place the cover down on a contaminated surface.
4. Inoculation loops or pipettes and fluids (such as saline or diluents), the culture medium, must first be sterilized.

5. Inoculating loop must be sterilized before and after use.
6. The pipettes you use to transfer cultures must be placed in disinfectant solution immediately after making the transfer.
7. For making smears on the slide from an agar slant, first transfer a loopful of water to the center of the microscope slide or cover slip.

Aseptic Transfer of Pure Culture

1. Heat the entire wire of the inoculation loop by the flame at the hottest part of a Bunsen burner.
2. Allow the loop to cool at least 15-30 seconds before is touches a culture.
3. Pick up a tube of culture with your free hand.
4. Remove the plug or cap of the tube with the fingers of the hand holding the inoculation loop.
5. Flame the mouth of the tube by monetarily rotating it within the upper cone of flame.
6. With the sterilized inoculating loop, remove a small amount of culture from the tube either by taking a loopful of broth or by gently touching the loop to the solid growth on a slant.
7. Flame the mouth of the tube again and replace the plug or cap, and return the tube to the test - tube rack.
8. Transfer the culture to the desired location on microscope slide or culture vessels without touching the culture to anything else during this transfer.
9. Flame your inoculating loop and set it down in a safe place after completing each aseptic transfer.

DETERMINATION OF pH AND MOISTURE PERCENTAGE

pH

Make suspension of 20 g of the biofertilizer sample into 50 ml of distilled water and shake for 2 hours. Filter this suspension and determine the pH of the filterate with the help of pH meter.

Moisture

Heat 10 g of sample for 12-16 hours in an air oven at 100-105°C. Cool in a desiccator and weigh. The loss in weight represents the moisture. Calculate the moisture percentage on air-dry weight basis, by multiplying the loss in weight by ten.

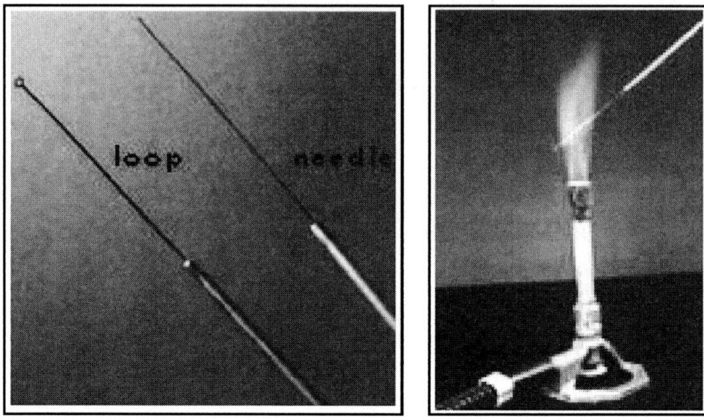

Flame the inoculating loop until the entire wire turns orange.

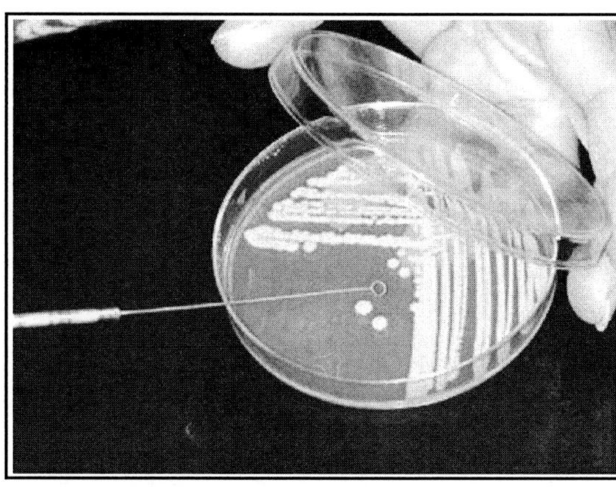

Aseptic transfer of pure colony

—•—

Chapter **22**

Equipments Used and their Principles

Horizontal Autoclave

Size: 0.9 x 0.9 x 1.8 m. It is an apparatus in which objects or materials may be sterilized by air-free saturated steam (under pressure) at temperature above 100°C. The word "Autoclave" is derived from French word where "Claves" means a Key i.e. device where automatic sterilization can be made. The autoclave is a cylindrical metallic vessel having a double wall. There are two types of autoclaves - Vertical and horizontal and their use is made as per requirement. Their Principle is as follows:

Water boils at about 100°C depending upon the vapour pressure of the atmosphere. If the atmosphere pressure is increased, the temperature will increase. Therefore, if the steam pressure inside the autoclave is increased to 15 psi (2 atm.), the temperature will rise to 121°C. This is sufficient to destroy vegetative cells and spores.

Operation

This type of sterilization involves an efficient transfer of heat since steam, which condenses on a cooler object, delivers a considerable quantity of latent heat. The Required temperatures for sterilization in this case are 115°C (240°F), 121°C (250°F), 135°C (275°F) etc. which can be achieved with steam under the pressure.

Precautions are to be Taken While Operating Autoclave

Evacuation of air from the chamber should be complete. If this has not been done, even through the pressure gauge indicates 15-psi pressure, the temperature will not reach 121°C, therefore, it is essential to measure sterilization time from the moment the thermometer reaches 121°C rather than from the time 15 psi is indicated.

If the materials to be sterilized does not contain liquids, the operating valve an be opened to release the pressure more rapidly. If liquids are in the chamber, sudden release of the pressure causes them to boil in the containers, wetting the plugs and blowing them out. It is always essential to let the pressure fall.

Horizontal Autoclave

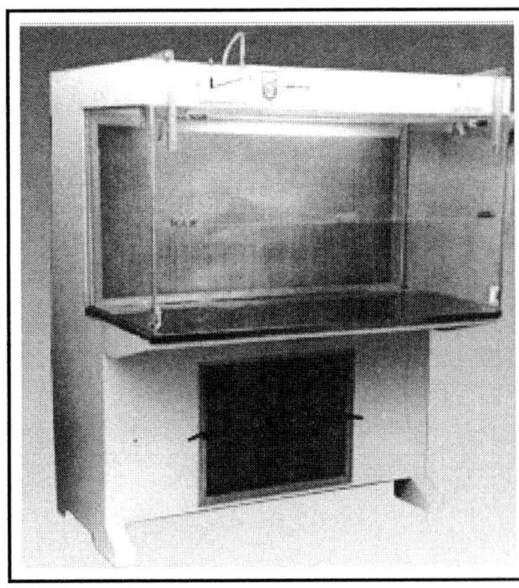

Laminar Air Flow Cabinet

Laminar Air Flow Cabinet (LAF Cabinet)

Laminar airflow cabinet is a structure in which a uniform flow of filtered air prevents settling of particles in the work area. At the time of isolation, inoculation or other microbiological work, use of this cabinet is recommended. The methods for control of airborne contamination rely on air flow in which the entire body of air within a confined area essentially moves with uniform velocity that cause an air flow to coverage, or to focus, onto the obstructing work objects such that purging of the low pressure zone with filtered air is achieved and airborne particulate contaminants are controlled.

Filters such as pre filters and HEPA (High Efficiency Particulate Air Filter) are used In LAF. The purpose of Prefilter (Coarse furnace filter Quality) is to screen out large particles before air is forced through a much finer HEPA filter which ultimately screens out particles than 0.3 μ. These filters consist of cellulose acetate plated around aluminum foil.

Microscope

Microscopes are used to produce a highly magnified image of a sample so that microorganisms can be seen, types of microscopes used in biofertilizer technology are described below:

Compound Microscope

It consists of two sets of lenses, the objective and the eyepiece. The eyes pieces further magnifies, while the chief function of the objective is to bring together the light rays coming from any point of the object, then unite them in a point of the image and to magnify image. Most microscopes manufactured have adopted 160mm as the standard length.

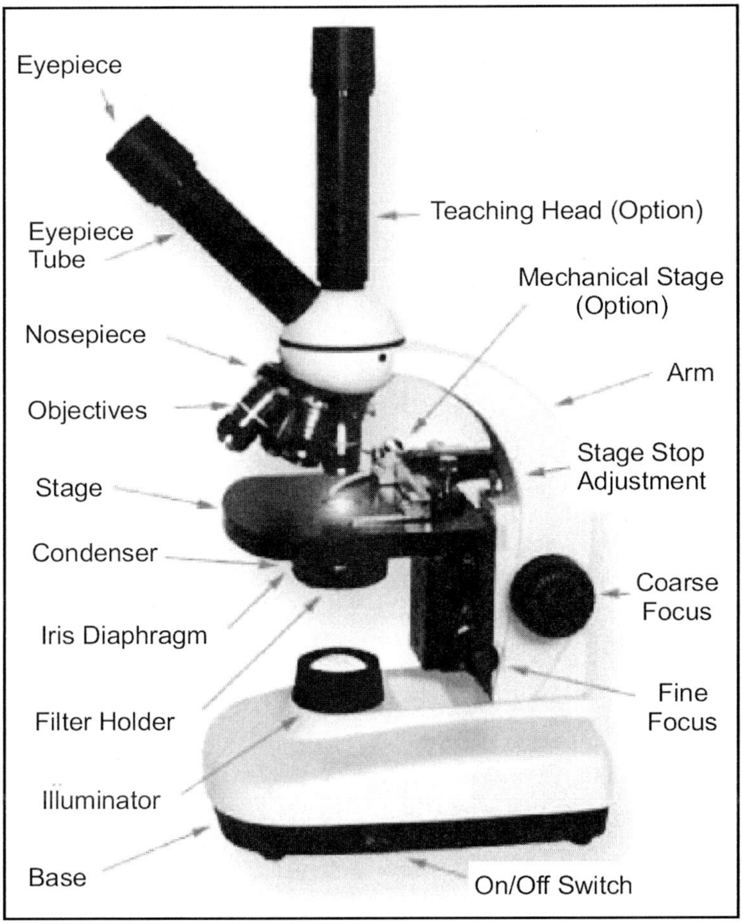

Compound Microscope

Phase Contrast Microscope

In this microscope, light rays passing through an object of high refractive index are retarded in comparison with light rays passing through a surrounding medium with a lower refractive index. The retardation or phase change for a given light ray is a function of the thickness and the refraction index of the material through which it passes. This microscope has become an extremely important tool in a microbiology laboratory because it intensifiers the contrast between translucent object in living, unstained specimens.

Double Decker Rotary Shaker

For shaking solutions in Erlenmeyer flasks. It should be compact bench model unit with double platform on ball bearings. Crankshaft should an orbital circular motion. The capacities can be either of the following.

- 1. 100 conical flasks of 1000 ml capacity on upper deck. The deck platform is fitted with adjustable rubber disks to hold the flasks.
- 2. 32 conical flasks of 5000 ml capacity on lower deck. Each flask holder is fitted with adjustable flasks neck holder.

The entire assembly should be mounted on a metal casting housing a 3 HP motor, speed controlled, having variable speed and heavy duty continuous rated. Entire assembly is ground tabled on the thick rubber pads so that the operation is smooth and noiseless.

Fermentor

100 to 1000 lit simple fermentor designed specially for the production of *Rhizobium/ Azotobacter / Acetobacter* etc. There should be good aeration, air outlet, sampling system, temperature, pH regulation and foam control.

Mechanical Powder Mixer

For mixing dry and finely ground powder with liquid to obtain 30-35% homogeneously mixed moist powder. Capacity 50 kg mixture per batch. Chamber made of 3 mm stainless steel sheet fitted horizontally on 4 mild steel legs welded firmly with the chamber. Chamber should be semicircular with stainless steel top which can slide easily in grieves made in top of the chamber in such a way that

the chamber is almost air tight when the lid is closed. Lid should be provided with 2 hoppers,1 large enough with 6″ to 8″ (15-20 cm) opening to allow easy movement of powder, other smaller one with funnel for liquid. Liquid hopper should be attached to horizontal perforated tube to allow even spreading of liquid. Chamber should be fitted with thick stainless steel shafting mounted on bearings and be fitted ribbon type arms or flanges in such a way that material added from one end can easily move automatically to the other end. There should be a 8″ (20 x 20 cm) wide opening at the bottom for withdrawing the mixed material. Shifting should be attached to speed reducing arrangement so as to provide 8-100 rpm and a motor of 2 hp phases. Reducing gears and motors deep fitted on another stand of MS attached to the main chamber.

Refrigerator

300 liters, two doors.

Refrigeration System

Shelf contained freeon 12 or freeon 22 or similar system. The compressor motor of high standard and effective, condenser copper tubing.

Power Supply

160- 260 V, 50 hz, construction as one piece cabinet, poly-urethane form insulation, effective door liner, magnetic door cabinet shelf, sturdy easy recoverable, rust resistant, temperature control switch, automatic defrost system.

BOD Incubator

Wheel mounted, triple walled, insulated with thick blanket of super glass wool, outer casting of mild steel attractively finished in white enameled stove paint, anodized interior aluminum walls allowing wide range of shelf positions. Full view inner glass door, closed on a resistance gasket. Lock and key arrangement in the double wall outer door. Fitted with door-operated lamp, temperature control range 5°C- 50°C ±0.5°C. Cooling coils and heating elements run all round the inner chamber.

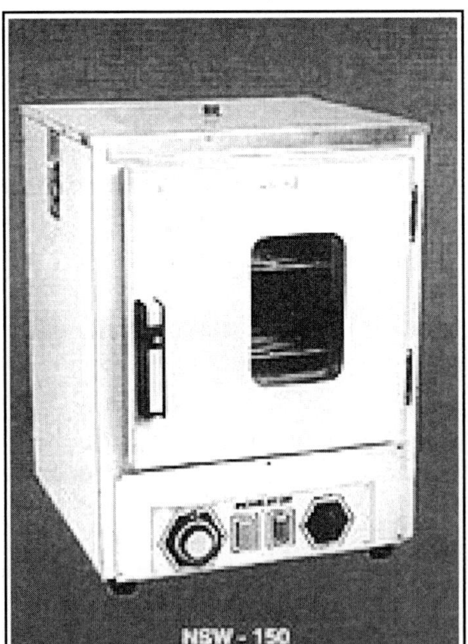

Incubator

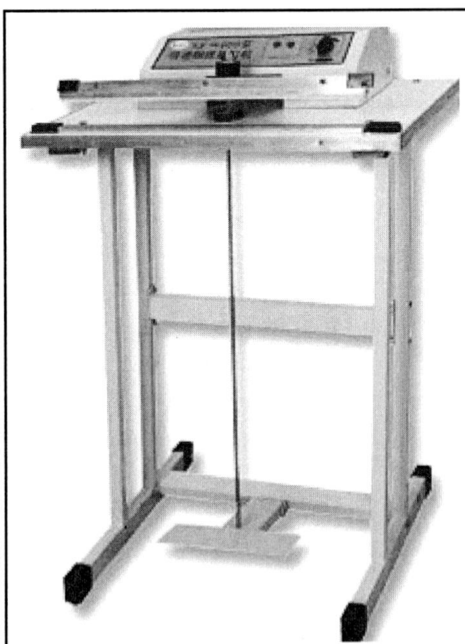

Plastic Bag sealing machine foot operated

Air circulation by a double shaft self-cooled fan. Safety thermostat be provided to switch off heaters in the event of failure of normal temperature control system. All controls at the top with electronic temperature indicator LED digital display. Separate indicator lamps for mains, heating and cooling with voltmeter. Unit be provided either built in or separate matching voltage stabilizer. Chamber size 900x700x650 mm (15cft) approximately.

Sealing Machine

Totally enclosed self supporting steel frame, dust and water protected, low vibration and noise rating, cross seal safety control with automatic reverse motion, safety switch for machine stop, when end of film is reached, open guard safety device, filling time is precisely synchronized with unit, Control of required sealing temperature, microprocessor controlled.

Laboratory Centrifuge

General-purpose laboratory centrifuge, maximum speed 6000 rpm with 3600-3800 g provided with graduated speed control knobs, times unit and speed meter. Rotation head 4x50 ml and 6x 16 ml. Whole unit housed in attractively finished rectangular box with stove spell check paint.

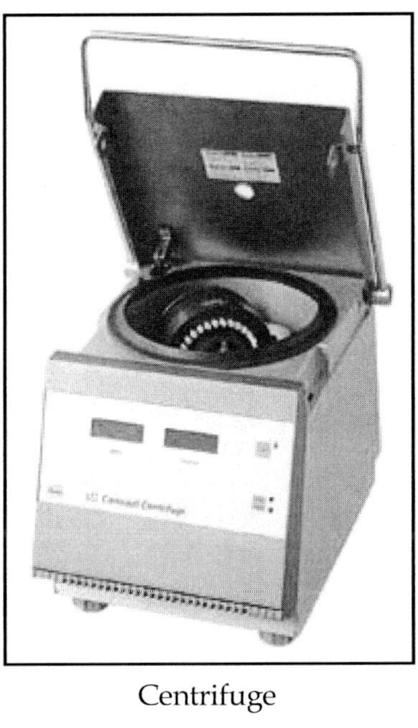

Centrifuge

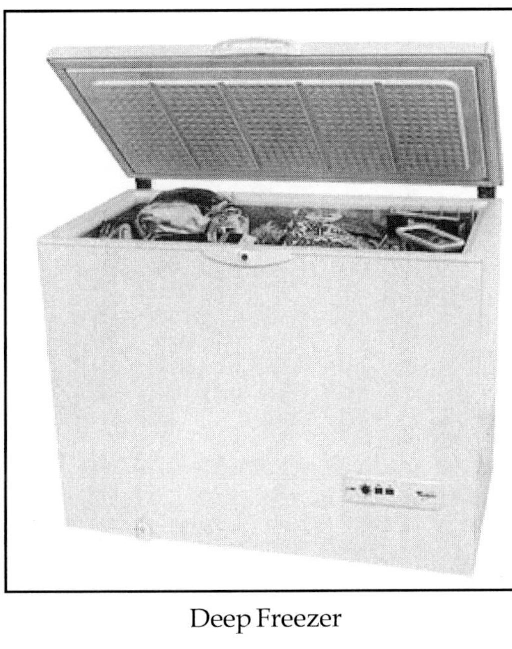

Deep Freezer

Deep Freezer (Horizontal)

Storage capacity 265 liters, inner size 1200x370x680 mm, outer size 1410x580x880 mm, temperature –15°C to –20°C, inside chamber made up of galvanized steel, insulation system, combination of glass wool and poly urethane foam of adequate thickness, tightly sealed door gasket.

Lyophilizer

Main supply 220-240 V, 50 hz, 1 pH; 110 V, 60 hz, 1 pH, power requirement 0.33 kw. Refrigerator size 0.33 hp, refrigerator heat extraction rate K cal/h at –35°C (at 20°C ambient 180,condensor ice capacity 1.2 kg in 12 h, 2.3 kg in 24 h and 4 kg total), condenser minimum temperature under vacuum –55°C, condenser surface fixed 0 15 m^2, volume 7.2 liter, height 332 mm, depth 690 mm, width 595 mm, weight 45 kg.

Hot Air Oven

Double walled with inner chamber made of stainless steel, white stove enameled paint. Insulation with special glass wool, heating system should be of memmert German type, temperature range 50°C to 250°C controllable by thermostats. Control panel be at bottom part with pilot lamps, on-off switch, hi-low switch, graduated temperature control knob and 0-60 minute timer. Chamber fitted with air circulating fan, chamber size 910 x 605 x 455 mm.

Lyophilizer

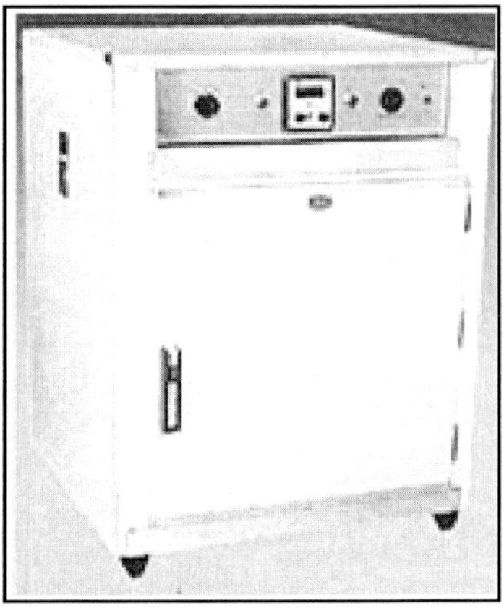

Hot Air Oven

Electronic Top Loading Balance

Capable of instant weighing in 2-3 seconds. Microprocessor based. Large bright LED read out, with auto calibration, cumulative memory, memory recall facility, one touch subtractive taring with audible beep. Capacity 0.001 to 100 g. Sensitivity + 0.001 g.

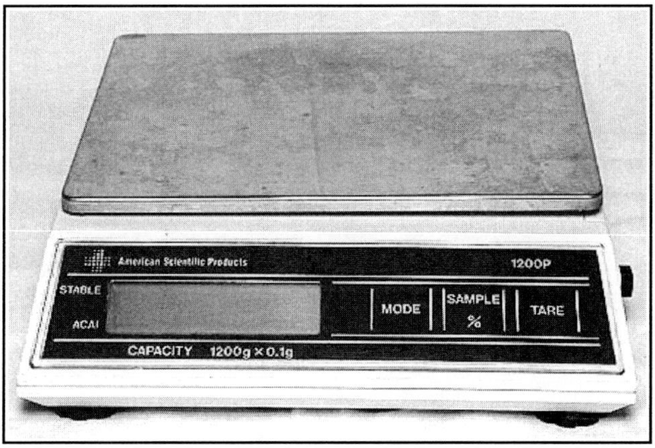

Electronic Top Loading Balance

Portable pH meter

pH meter

Microprocessor based with bright LED display of pH and temperature simultaneously. PH range 0-14, temperature range 0-100 C.

Electronic Colony Counter

Should provide uniform glass free illumination. Entire area covered with a convex lens, counting plate with standard welfhugle sealing. Fitted with automatic digital display, on-off switch and probe of counting.

Colony Counter

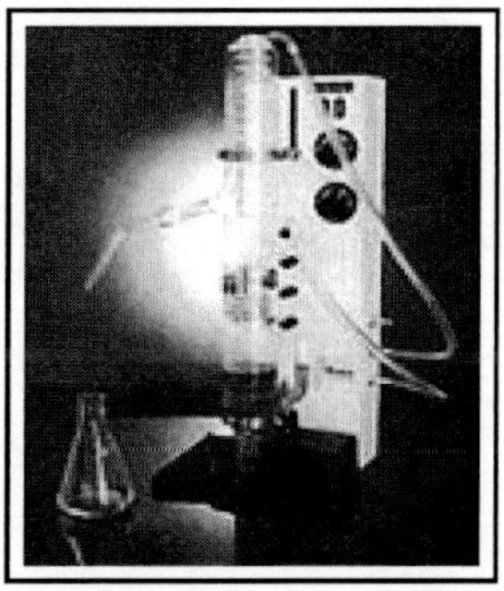

Glass Distillation Apparatus

Glass Distillation Apparatus

Automatic electrically heated, heater embedded in glass tube complete with metallic stand, plug, and board etc, double stage, 5-10 lit/hour capacity, Preferably quartz glass.

—•—

Chapter 23

Research and Development at Anand Agricultural University

BIOFERTILIZER

Recommendation for Farming Community

In order to develop a low input, eco-friendly technology, following recommendations involving use of ***Azolla, Azotobacter, Azospirillum,*** **Blue green algae and Phosphate culture** in different crops have been released for farming community by Anand Agricultural University.

(A) Nitrogen Fixer

Azotobacter chroococcum ABA-1 , *Acetobacter diazotrophicus* and *Azospirillum lipoferum* ASA-1

(a) Pearl Millet

In Middle Gujarat, seed treatment of **Bajra with *Azotobacter* ABA-1** or ***Azospirillum ASA*-1** culture @ 200 g/10 kg seed having 10^8 viable cells/g, proved effective in saving half of the recommended dose of nitrogen i.e. 40 kg N/ha with I.C.B.R. 1:10.22 and 1:9.12, respectively (1991).

Farmers of Middle Gujarat Agro climatic Zone growing summer transplanted **Pearl millet are** advised to dip seedlings into suspension of ***Azospirillum*** cultures such as **GMF 1601** (ICBR 1:210) or **ASA-1** (ICBR 1:162) or **BF 0002** (ICBR 1:114) (4 packets each of 250 g/ha, containing 10^8 CFU/g carrier, suspended in 15 liters of water) for 15-

20 minutes just before transplanting and to fertilize with 40 kg N/ha to get maximum yield. (2001).

(b) Sorghum

Sorghum seeds inoculated with 'Endogonous' ***Azotobacter*** culture gave net I.C.B.R. of 1:182 and 1:95 for grain and fodder yield, respectively (1987).

For obtaining higher **Sorghum** grain and fodder yields, seed inoculation either with ***Azospirillum* ASA- 1** (ICBR 1:10) or ***Azotobacter* ABA- 1** (I.C.B.R. 1:9.46) each having 10^8 viable cells/g (200 g culture/10 kg seeds) along with the recommended dose of 40 kg N/ha is recommended for marginal farmers of Gujarat (1994).

(c) Rice

Farmers of South Gujarat heavy rainfall zone are advised to adopt the techniques of seedling root dipping for 15 minutes in one of the following ***Azotobacter* or *Azospirillum*** culture suspension before transplanting and use only 60 kg N/ha in **Summer Rice** GR 11 instead of 120 kg N/ha (1992).

Culture	I.C.B.R.	Culture	I.C.B.R.
Azotobacter ABA 1	1:28	*Azospirillum* ASA 1	1:26
Azotobacter BF 1007	1:24	*Azospirillum* 0013	1:23
Azotobacter BF 1005	1:23	*Azotobacter* BF 0009	1:23

Farmers of agro climatic zone III are advised for dipping of rice seedling roots in biofertilizer suspension of ***Azotobacter* BF 1013** (ICBR 1: 3.45 or ***Azospirillum* ASA- 1** (ICBR 1:3.4) having $1x10^8$ cells/ g for 15 minutes and adding remaining biofertilizer suspension to the soil before transplanting along with 75 kg N/ha to save **25 kg N/ha.** (1997).

(d) Finger Millet

In tribal area of Waghai (Dang Dist.) 20 kg N/ha in **Finger Millet** can be saved through mere seed coating with ***Azospirillum*** culture (1990).

(e) Maize

Farmers of Middle Gujarat Agro climatic Zone are advised to treat **Maize** seeds with ***Azotobacter* ABA- 1** (ICBR 1:173) or ***Azospirillum***

ASA-1 (ICBR 1:143) @ 25 g/ kg seeds (10^8 cfu/g) to save 20 kg N/ha along with 30 kg P_2O_5/ha as basal dose and seeding fortnight of June (onset of monsoon) to escape from maydis leaf blight, brown stripe downy mildew and dead heart diseases for getting higher grain and fodder yields. (2001).

(f) Rajagara

For getting maximum yield and net realization of Rs.13410/ha, the farmers are advised to **Rajagara** seeds with ***Azotobacter*** **ABA- 1** culture (having 10^8 viable cells/g) @ 200 g/kg seed before sowing.(ICBR 1:204) (1997).

(g) Wheat

For the efficient use of the fertilizers the farmers growing **Wheat** under **irrigated** condition of North Gujarat Agro climatic zone AES-4 are advised to treat wheat seeds with ***Azotobacter*** **ABA –1** and **PBA-4 (*Bacillus circulans*)** (15 pkt/120 kg seed/ha) before seeding followed by 75 kg N, 30 kg P_2O_5 as basal dose 37.5 kg/ha N as top dressing at first node stage i.e. 35 DAS, to economize N and P_2O_5 to the extent of 25 % and 50 %, respectively. (2003).

The farmers of Middle Gujarat Agro climatic Zone (AES-2) growing **Wheat - Fodder Sorghum** (summer) and interested in sustainable agriculture through integrated nutrient management are advised to use 75% recommended N (90 kg/ha) in 2 equal splits) with 12.5% N through FYM or 50% N (60 kg/ha in 2 splits) with 25% N through FYM in combination with ***Azotobacter*** **(ABA –1 or ABA-2) or *Azospirillum*** **(ASA-1** or **ASA-5)** as seed (GW-503) treatment (with 30-45 kg P_2O_5 as basal application), followed by 100% recommended fertilizer (80-40-0 NPK) to the summer fodder sorghum (S-1049) to get crop yields equivalent to application of 100% recommended fertilizer to both the crops for sustaining productivity of soils. (2003).

(h) Pigeon pea

Farmers of middle Gujarat Zone-III (ASE-II) growing **Pigeon Pea** (BDN-2) in kharif are advised to use

(i) RDF + Rhizobium (RBA-5)+Azotobacter (ABA-1) OR

(ii) RDF + Rhizobium (RBA-5)+Azotobacter (ABA-1) + phosphorous solubilizer (PBA-22) through either seed treatment (30 g/kg seed) or soil application (5 kg/ha) for getting higher yield and net return. (2006).

(h) Forage crops

(i) Sorghum

In Middle Gujarat, seed treatment of sorghum for fodder with ***Azotobacter* ABA- 1** or ***Azospirillum* ASA-1** culture @ 200 g/10 kg seed having 10^8 viable cells/g, proved effective in saving half of the recommended dose of nitrogen i.e. 40 kg N/ha with I.C.B.R. 1:7.80 and 1:15.57, respectively (1991).

The farmers of middle Gujarat Agro climatic Zone (AES-2) growing fodder sorghum variety SSG.59-3 are advised to treat the seed with ***Azotobacter*** for getting higher dry matter yield and crude protein production. Nitrogen should also be applied @ 50 kg/ha in two equal splits, first as basal and the second 30 DAS for one cutting management and for two cuttings management top dressing should be immediately after first cutting. (1991).

The farmers of Middle Gujarat Agro climatic Zone (AES-2) growing **wheat - fodder sorghum** (summer) and interested in sustainable agriculture through integrated nutrient management are advised to use 75% recommended N (90 kg/ha) in 2 equal splits) with 12.5% N through FYM or 50% N (60 kg/ha in 2 splits) with 25% N through FYM in combination with *Azotobacter* (ABA -1 or ABA-2) or *Azospirillum* (ASA-1 or ASA-5) as seed (GW-503) treatment (with 30-45 kg P_2O_5 as basal application), followed by 100% recommended fertilizer (80-40-0 NPK) to the summer fodder sorghum (S-1049) to get crop yields equivalent to application of 100% recommended fertilizer to both the crops for sustaining productivity of soils (2003).

(ii) Guinea grass

The farmers of middle Gujarat Agro climatic zone growing guinea grass are advised to give *Azospirillum* treatment to guinea grass seedlings and apply 30 kg N/ha during each cutting for getting higher green forage, dry matter and crude protein yield (1992).

(i) Cash Crops Sugarcane

The farmers of middle Gujarat Agro-climatic Zone-III (ASE-II) growing sugarcane (Co-6304) are advised to fertilize the crop with 150 kg N + 125 kg P_2O_5 +125 kg K_2O per ha and use sugarcane sets for planting after a treatment of *Acetobacter diazotrophicus* ACG-2 (containing 10^8 CFU/g carrier) @ 4 kg/ha to save 100 kg N/ha and get higher cane yield (14%) and net return (2006).

(B) Phosphate Culture as Phosphate Solubilizer

(a) Pigeon Pea

Farmers of South Gujarat Zone (AES V) are advised to coat **Pigeon pea** seeds with PSM strain PBA-22 ***Torulospora globosa*** (ICBR 1:513) or PBA-13 ***Bacillus coagulans*** (ICBR 1:379) each having 10^8 CFU/g carrier @ 30 g culture/kg seed before seeding to save **40 kg P_2O_5/ha** and get maximum green pod yield (1998).

Farmers of middle Gujarat Zone-III (ASE-II) growing pigeon pea (BDN-2) in *kharif* are advised to use

(i) RDF + *Rhizobium* (RBA-5)+*Azotobacte*r (ABA-1) OR

(ii) RDF + *Rhizobium* (RBA-5)+*Azotobacter* (ABA-1) + phosphorous solubilizer (PBA-22) through either seed treatment (30 g/kg seed) or soil application (5 kg/ha) for getting higher yield and net return. (2006).

(b) Sorghum

Farmers of South Gujarat Agro climatic Zone AES II growing **Sorghum** are advised to coat seeds with PSM strain PBA-16 ***Bacillus coagulans*** (ICBR 1:382) having 10^8 CFU/g carrier @ 30 g culture/kg seed before seeding to save **40 kg P_2O_5/ha** and to get higher grain and stover yield (1998).

Farmers of Middle Gujarat Agro climatic Zone growing fodder **Sorghum** are advised to coat seeds with phosphate culture such as PBA-22 ***Torulospora globosa*** (ICBR 1:24) or PBA 12 ***Bacillus brevis*** (ICBR 1:24) or PBA 16 *Bacillus coagulans* (ICBR 1:22) @ 30 g/ kg seeds (10^8 cfu/g) to get maximum fodder yield (2001).

(c) Wheat

Farmers of Bhal and Coastal Agro climatic Zone growing **rain fed wheat** are advised to coat seeds with phosphate culture, such as ***Bacillus coagulans* PBA-14** or **PBA-13 or *Bacillus brevis* PBA-12** @ 6 packets /ha containing 108 CFU/g carrier and application of 20 kg P_2O_5 /ha as basal dose to get maximum yield and net realization (2003).

(d) Cowpea

Farmers of South Gujarat Agro-climatic Zone growing **Cowpea** are advised to coat seeds with phosphate solubilizing culture such as PBN-1 (ICBR 1:522) or PBA -14 (ICBR 1: 362) or PBA- 4 (ICBR 1: 346)

@ 30 g culture/kg seed containing 10^8 cfu/g at seeding to save recommended dose of P_2O_5/ha and get higher yield. (2002).

(e) Uridbean

Farmers of South Gujarat Agro-climatic Zone growing **Uridbean** are advised to coat seeds with phosphate solubilizing culture such as PBA-17 (ICBR 1:453) or PBN -2 (ICBR 1: 312) or PBA- 16 (ICBR 1: 244) @ 30 g culture / kg seed containing 10^8 cfu/g at seeding to save recommended dose of P_2O_5/ha and get higher yield (2002).

(f) Groundnut

Farmers of South Saurashtra Agro-climatic Zone growing are advised to apply PSM cultures either PBA-22 **(*Torulospora globosa*)** (ICBR 1: 8.62) or PBA- 17 ***(Bacillus coagulans*)** (ICBR 1: 2.66) (10^8 cfu/g) as seed treatment @ 30 g culture / kg seeds in **Groundnut** crop in place of phosphatic fertilizers to increase yield by 26.7% and 19% respectively (2003).

(g) Sesamum

The farmers of AES –IX of north Saurashtra Agro climatic zone growing **sesamum** are advised to apply 5 ton of FYM or 1 ton castor cake /ha with 50% recommended fertilizer (25:12.5:0 NPK kg/ha) along with seed treatment viz. ***Azospirillum*** and **PSB** for obtaining maximum seed yield of **Sesamum** (1998).

(h) Maize

Farmers of middle Gujarat Agro-climatic Zone-III (AES-III), growing kharif maize (GM-4) are advised to inoculate seeds with *Azospirillum lipoferum* ASA-1 and *Torulospora globosa*_PBA-22 or *Bacillus coagulans* PBA 16 having 10^8 CFU/g carrier @ 30 g culture per seed before sowing along with soil application of 40 kg N/ha and 20 kg P_2O_5/ha to save 20 kg each of N and P along higher yield and economic return (2006).

(i) Pearl millet

Farmers of South Saurashtra Agro-climatic Zone growing **bajra** are advised to apply PSM cultures either **PBA-16 (*Bacillus coagulans*)** (ICBR 1: 61.47) or **PBA- 12 (*Bacillus brevis*)** (ICBR 1: 51.88) (10^8 cfu/g) as seed treatment @ 30 g culture / kg seeds in **Bajra** crop in place of phosphatic fertilizers to increase yield by 26.5% and 22.6%, respectively (2003).

(j) Chickpea

The farmers of Middle Gujarat Agro-climatic zone -III cultivating **gram** crop (GG-2) are advised to apply 75 % of recommended phosphorus fertilizer @ 30 kg/ ha and phophobacteria culture with adequate organic manure as a basal treatment in conjunction with irrigation in two stages (i.e. branching and pod formation) for getting higher yield as well as higher net profit (2005).

MICROBIAL PESTICIDE

Recommendations for Scientific Community

1 It is recommended that application of **bio-nematicide, *Paecilomyces lilacinus*** spore dust with substrate carrier @ 50 kg/ha (10^9 conidia/g product) was found effective for management of *Meloidogyne javanica* (pt.2), which was at par with carbofuran @ 1kg/ha (1999).

2 It is recommended that application of **bio-nematicide, *Paecilomyces lilacinus*** @ 25kg/ha (granules having spore load 5×10^8 conidia/g) with phenamiphos @ 1kg/ha was found most effective for management of root-knot nematodes, *Meloidogyne spp.* with highest yield of cotton. Phenamiphos alone and *P. lilacinus* with Carbofuran @ 1kg/ha were next effective treatments. *P. lilacinus alone* was at par with carbofuran , the popular nematicide (2000).

3 **Entomopathogenic nematode**, ***Steinernema glaseri*** with anti desiccant was found effective against *Plutella xylostella* on cress inducing 59.1% reduction in larval population. Biopesticide treated check, *Bacillus thuringiensis*, treatment remained most effective giving 87.5% reduction of larval population with highest cress seed yield followed by monocrotophos treated check (2000).

—•—

Chapter 24

Opportunities in Agrobiotech Industries

Biofertilizers and Biopesticides

Biofertlizers are carrier based microbial inoculants containing cells of efficient strains of microorganisms. They are used for enhancing the productivity of the soil. They fix atmospheric nitrogen and solubilize phosphorus. They also stimulate plant growth through production of plant growth promoting substances.

The low cost, high efficiency technology of biofertiliser application has an important role to play in increasing agricultural production without rapid depletion of non-renewable energy. Role of biofertlizers in the agriculture enterprise becomes more important as it economizes the fertilizer application to crops. Application of biofertilizer has thus, become an integral component of nutrient management system. Apart from this, Biofertlizers play important role in increasing the agricultural production, a need of the day.

Staff required for commercal Unit (150 mt/annum)

Project Manager	1
Senior Microbiologist	1
Assistant Microbiologist	2
Senior Research Assistant	2
Junior Research Assistant	3
Agricultural Assistant cum field man	2
Laburers	5

Production

The unit will produce the following biofertilizers and microbial pesticides.

1. *Azospirillum* for cereal crop etc.
2. *Rhizobium* for pulses.
3. *Acetobacter* for sugarcane.
4. Phospho bacterium for all crops.
5. *Azotobacter* for all crops.
6. *Trichoderma viride.*
7. *Pseudomonas.*
8. *Bacillius thuringienrs*
9. *Entomopathogenic muscardine* fungi

CURRENT ANNUAL BIOFERTILIZER PRODUCTION AND DEMAND IN INDIA (TON/YEAR) APPROXIMATELY

Market Potential

Biofertilizer demand all over India	12,43,940 M.T.
Biofertilizer demand in Gujarat	50,00 M.T.
Bio fertilizer produced in Gujarat at present (Including government & Private institution)	1250 M.T.

The gap is due to lack of awareness among the farmers.

Advantage

- This unit can give employment to 10-15 people, which depends upon the capacity of production.
- As the awareness for organic farming among the farmers is increasing, the demand for bio-products also increasing day by day.
- Minimize the use of chemical fertilizers and pesticides.
- Increasing crop yield by using the biofertilizer.
- Produce the quality product, which will give better price to farmers.
- To save the environment.
- Increase the soil fertility.

Disadvantage

- The awareness among the farmers using the biofertilizer is low.
- The shelf of life Biofertilizer is very short (6 month). We cannot keep more stock and sale in season.
- For meeting sudden demand in season, we have to increase the capacity of the unit at 1 point by investing more money in plant and machinery.
- We cannot guess exact demand because it demands upon rain in the area.

Remedies

- The state government should encourage the biofertilizer unit by purchasing their production for the demand of their departments.
- The state government can reduce and eventually top the commercial production of biofertilizer by their agriculture department.
- The service of the staff can be utilized for creating awareness among the farmers.
- The government can give guidance and publicity and conduct trials in the field of farmers to increase awareness.
- Conduct R&D in the biofertilizer related aspects.
- Check the quality and price of unit holders for getting good quality biofertilizers.
- The state government encourages the entrepreneurs to start biofertilizers unit at least 1 or 2 unit in each districts to avoid transportation of products for long distance so that the farmers can get quality products at reasonable price at required time.

Area of Marketing

- Sugar mills
- Tea and coffee and other plantation crops
- Agriculture department
- Forest department
- Sericulture department
- Directly to farmers through fertilizer dealers.

—•—

Bibliography

FURTHER READING: Books/ Reviews

Biofertilizers

Biofertilizer Technolgy, Marketing and Usage (1995) Dr. M. R. Motsara, Dr. P Bhattacharyya and Dr. Beena Srivastava, FDCO, New Delhi,India.

General Microbiology (1986) Roger Y. Stanier, John L. Ingraham, Mark L. Wheelis and Page R. Painter (Eds) The Macmillan Press Ltd., London.

Fertilizers, organic manure, recyclable waste and biofertilizers (1992) H.L.S Tandon. FDCO, New Delhi,India.

Handbook of Organic farming and biofertilizers (2006) A.C. Gaur (Ed) Ambica Book Agency, Jaipur.

Hndbook on biofertilizers (2005) L.L.Somani (Ed) Agrotech Publishing Academy, India.

Microbiology (1993) Michael J Pelczar, Tata McGraw Hill, UK.

Soil Microbiology (1977) N.S.Subba Rao, Oxford and IBH Publising Co. Pvt. Ltd., India.

Soil Microbiology: An exploratory approach (1999) Mark S. Coyne (Ed) Delmar Publishers, UK.

Websites:

http://nabard.org

http://www.gsfclimited.com

Microbial pesticides

Biological control of plant parasitic nematodes, G.R. Stirling (ed.), CAB Publication, Wallingford, UK, (1991), pp. 281

Biological Control of Root-knot Nematodes in India. Vyas, R. V. and Patel, D. J. Annual Review of Plant Pathology (2002) (1) pp. 137-154.

Microbial control of insects and mites. (Eds) H.D. Burges and N.W. Hussey, Academic Press London and New York (1971), pp. 876

Microbial control and pest management. (Ed) S. Jayaraj, Tamil Nadu Agricultural University, Coimbatore 641 003, India (1985), 278pp.

Microbial control of pests and plant diseases 1970-1980. (Ed.) H.D. Burges, Academic Press London and New York (1981), pp. 949

Nematode Bicontrol, M.S. Jairajpuri, M. M. Alam and I. Ahmad (Eds.), CBS publisher, Delhi, India, (1990) pp.152

Literature cited

Anonymous (1983). Annual Report, Fourth All India workshop on Biological Control, Banglore, India. 22nd -24th Feb. 1983, p. 217.

Anonymous. (1997). Phase I Report of Project 'Nematodes of Agriculture', Dept of Nematology, Guj. Agril. Univ., Anand, 40 pp.

Anonymous. (1999). Phase II Report of Project on 'Nematodes of Agriculture', Dept. of Nematology, Guj. Agril. Univ., Anand, 32 pp.

Anonymous. (2000). Final Technical Report of Project ' Management of *H. armigera* using Entomopathogenic nematodes'. Dept. of Nematology, Guj. Agril. Univ., Anand, 34pp.

Anonymous. (2001). A potential entomopathogenic nematode. In: Research Highlights, GAU News, 4(1): 3.

Anonymous (2005). An introduction to biofertilizers and microbial pesticides, Training Programme on Production Technology and Quality control of Biofertilizers, October 24–29, 05, at AAU, Anand, India.

Bagyaraj, D. J., Manjunath, A. and Reddy, D. D. R. (1979). Interaction of vesicular arbuscular mycorrhiza with root-knot nematodes in tomato. Plant and Soil Biol. 51 : 397-403.

Bhatti, D.S. and Walia, R.K. (1992). Nematode Pests of Crops, CBS Publisher, Delhi, India. 381pp.

Cabanillas H.E., Wright, R.J. and Vyas, R.V. (2005). Entomopathogenic Nematodes: Cereal, Fiber, Medicinal and Oilseed Crops Applications. *In*: Nematodes as Biocontrol Agents (Eds. Parwinder Grewal, Ralf Ehlers and David Shapiro-Ilan), CABI, UK.USA, 265-280pp .

Carlo, M. L.; William, C. Y.; Seth, Y. Y.; Donald, L. H. and Donald, L. B.(1976). Laboratory and field persistence of new commercial formulations of the *Heliothis* Nucleopolydedrosis Virus, *Baculovirus heliothis* J. Econ. Entomol., 69 (2) : 233-236.

Chahal, P.P.K. and Chahal, V. P. S. (1996). Biological control of root-knot nematode on brinjal (*Solanum melongena* L.) with *Azotobacter chroococcum*. In: M. A. Maqbool, A. M. Golden, M .A. Ghaffar and L. R. Krusberg (eds.), Advances in Plant Nematology. National Nematological Centre, Karchi, Pakistan: pp 257-263.

Charati, S.N., Mali, J. B.and Pawar, N. B. (1998). J. Maharashtra Agric. Univ., 23 (3): 304-305

Chattopadhyay, C., Meena, P. D. and Sudhir Kumar (2002). J. Mycol. Pl. Patho., 28(2) : 194-200

Dasgupta, D.R. and Gaur, H.S. (1986). The root-knot nematodes, *Meloidogyne* spp. in India. *In:* G. Swarup and D.R. Dasgupta (eds.), Plant Parasitic Nematodes of India Problems and Prospects. Director, IARI (Allied Publisher), New Delhi, India : pp 139-171.

Dubey, S.C.(2000). J. Mycol. Pl. Patho., 30 (1) : 80-90

Eapen, S. J. and Venugopal, M. N. (1995). Field evaluation of *P. lilacinus* and *Trichoderma* spp. in Cardamom nurseries for the control of root-knot nematodes and rhizome rot disease. *In*: Proc. of National symposium on Nematode Problems of India- An appraisal of the nematode management with eco-friendly approaches and biocomponents. March 24-26, 1995, IARI, New Delhi, India. pp 43 (Abrs.).

Ethiraju, S.; Rabindra, R. T. and Jayaraj, S. (1988). Laboratory evaluation of certain formulations of nuclear polyderosis virus against the larvae of Heliothis armigera. J. Biol. Cont., 2 (1): 21-25.

Fan Xiu Hua; Feng Shu Liang; Fu Yun Qin; Wang Rong Yong and Hu Ming Jun (1999). Field studies on the occurance and epidemology of entomopathogenic diseases infecting *Helicoverpa armigera* in Hebei. Chinese Journal of Biological Control 15 (1): 27-30.

Ferron, P. (1978). Biological control of insect pests by entomogenous fungi. Ann. Rew. Ent., 23: 409 – 442.

Ganguly, S., Singh, M., Lal M., Singh, L.K., Vyas, R.V. and Patel, D.J. (2002). A report on the identification of *Steinernema riobrave* Cabanillas, Poinar & Rqulston, 1994 from Anand, Gujarat. Paper presented in National symposium : Biodiversity and Management of Nematodes in Cropping Systems for Sustainable Agriculture at Agriculture Res. Station, Durgapura, Jaipur 302 015 on Nov. 11-13, 2002. : P.77(Abst.)

Gavas, R. and Ganguly, S. (2002). Effect of soil depth on infectivity of *S. thermophilum* Ganguly & Singh, 2000. Proceedings of National Symposium on Biodiversity and Management of Nematodes in Cropping Systems for Sustainable Agriculture, organized by NSI, ICAR and RAU at Dept. of Nematology, ARS, Durgapura, Jaipur on Nov. 11-13, 2002, : 77 (Abstr.).

Gokte, N. and Swarup, G. (1988). On the potential of some bacterial biocides against root-knot nematodes and cyst nematodes. Indian J. Nematol. 18(1):152-153.

Gopal Krishnan, c. and Naranganan, K. (1990). Studies on the dose mortality relationship between the entomofungal pathogen *Beauveria bassiana* (Bels.) Vulleimin and *Heliothis armigera* (Huber) (Lepidoptera: Noctudiae). J. Biol. Cont. 4(2):112-115.

Gopalakrishnan, C. and Narayan, K. (1988). Occurance of two entomofungal pathogens, *Metarhizium anisopliae* (Metchnikoff) Sorokin var. minor Tulloch and *Nomuraea rileyi* (Farlow) Samson on *Heliothis armigera* Hubner (Noctuidae: Lepidoptera), Curr. Sci. (57) : 867-868.

Gowda, J. and Ravi prasad, T.W. (1992). Occurrence of *Beauveria bassiana* (Bab.) vuils.on the Redgram pod borer *Heliothis armigera* Hub. J. Biol. Cont. 6(1): 48-49.

Hershman, T. (1995). New Bionematicide in Israel. Israel Agritec. Foc. 4(7): 14

Hussaini, S.S.; Singh, S. P.; Parthasarathy, R. and Shakeela, V. (2001). Comparison of bioassays to measure the infectivity of some isolates of EPNs against *Agrotis ipsilon* (Hufnagel) and *A. segetum* (Schiff).Curr. Nematol., 12(1, 2): 63-68.

Ignoffo. C.M; Garcia, C; Alyoshina, O.A. and Lappa, N.V. (1979). Laboratory and field studies with Baverin: a myco insecticidalpreparation of Beauveria bassiana product in the soviet union. J. Econ. Entomol. 72 (4): 562 – 565.

Jayraj, S. and Mani, A. (1988). Biocontrol of *Meloidogyne javanica* with the bacterial spore parasite Pasteuria penetrans. Int. Nematol. Network Newsl. 5(1): 3-4.

Jonathan, E. I., Padhmanabhan, D. and Ayyamperumal, A. (1995). Biological control of root-knot nematode on betelvine by *Paecilomyces lilacinus.* Nematol. Medit. 23 : 191-193.

Joseph rajkumar A. and C. V. Sivakumar (1996). Storage life and virulence of a native *Steinernema* spp. multiplied *in vitro*. Indian J. Ent., 58 (4): 318-321.

Jun, M.A. and Milner, R.J. (2000). Analysis and modeling of time-dose-mortality of *metarhizium anisopliae* infecting larvae of *Plutella xylostella*. Chinese Journal of Bio Control. 16 (1): 15-18.

Khan, M. R. and Goswami, B. K. (1999). Nematicidal effect of culture filtrates of Paecilomyces lilacinus isolates on *Meloidogyne incognita*. Indian J. Nematol.29(2) : 145-148.

Khan, M. R. and Goswami, B. K. (2000). In vitro evaluation of Indian isolates of *Paecilomyces lilacinus* against *Meloidogyne incognita* eggs. Indian J. Nematol. : 96-97.

Koshi, P. K. and Gulsar Banu, J. (1998). Use of fungal and bacterial antagonists for the biological control of nematodes in plantation crops. *In*: S. P. Singh and S. S. Hussaini (eds.) Biological Suppression of Plant disease, Phytoparasitic nematodes and Weeds. Project Directorate of Biological Control, Bellary Road, Bangalore 560 024, India : pp 154-163.

Maghodia, A B.; Vyas R. V.; Patel, Biren A. and Patel, D. J. (2003). Antagonistic effect of native *Steinernema riobrave* against root-knot nematodes *Meloidogyne incognita* and *Meloidogyne javanica* pt. 2. J. Mycol. Pl. Pathol., 33 (3) : 487

Maheshwari, U.T. and Mani, A. (1988). Combined efficacy of *Pasteuria penetrans* and *Paecilomyces lilacinus* on the biological control of *Meloidogyne javanica* on tomato. Int. Nematol. Network Newsl. 5:10-11.

Malathi, S. and Sriramilu, S. (2000). Laboratory efficacy of biotic insecticides against lepidopterous pests feed on treated cabbage leaves. Shashpa, 7 (1) : 63-66.

Mathasoliya, J. B., Maghodia A. B. and R. V. Vyas. (2004). Efficacy of *Steinernema riobrave* against *Agrotis ipsilon* Hufnagel (Lepidoptera: Noctuidae) on potato. Indian Journal of Nematology. 34(2): 177-79

Mishra, M. P.; Pawar, A. D. and Ram, N. (1991). Use of NPV in management of the insect pest, *Heliothis armigera* (Hubner) in gram. J. Andaman Sci. Assoc., 7 (1& 2): 75-78.

Nagesh, M., Parvatha Reddy, P. and Rao, M. S. (1996). Management of *Meloidogyne incognita* on *Crossandra undulaefolia* using arbuscular mycorrhiza *Glomus mosseae* and oil cakes. Mycorrhiza 9 : 12-14.

Narayanan, K. (1979). Studies on nuclear polydedrosis virus of gram caterpiller, *Heliothis armigera* (Hub.) (Noctuidae : Lepidoptera). Ph. D. thesis, Tamil Nadu Agricultural University, Coimbatore, p:109.

Narayanan, K. (1980). Field evaluation of a nuclear polyhedrosis virus of *Heliothis armigera* (Hub.) on chickpea, *Cicer arietinum* L. Proceedings workshop on biological control of *Heliothis sp*. Dept. of Primary Industries. 23-25. Sept. Toowomba, Queensland, Australia.

N. R. (1990). Field efficacy of nuclear polyhedrosis virus against *Heliothis* (Lepidoptera: Noctuidae) on gram (Cicer *arietium*) in Maharashtra. Indian J. Agric. Sci., 60 (4): 287-289.

Pandey, K. K. and Upadhyay, J.P.(1999). J. Mycol. Pl. Patho., 29 (2) :214-216

Parvatha Reddy, P. and Nagesh, M. (1998). Fungal and bacterial antagonists for biological suppression of plant parasitic nematodes on horticultural crops *In*: S. P. Singh and Hassaini S. S. (eds.) Biological suppression of Plant diseases, Phytoparasitic nematodes and Weeds. Project Directorate of Biological Control, Bellary Road, Bangalore 560 024, India. pp 138-153.

Patel, D.J., Vyas, R.V., Patel, B. A. and Patel, R. S. (1995). Bioefficacy of *Paecilomyces lilacinus* (Thom.) Samson in controlling *Meloidogyne javanica* (Pathotype 2) on groundnut. Int. Arachis News Lett. 15 : 46-47.

Patel, M.C. and Vyas, R.V. (1995). Efficacy of *Steinernema glaseri* against *Helicoverpa armigera* on chickpea in pots. International Chickpea & Pigeonpea News Letter, 2: 39-40.

Patel, N. S. (1999). Use of native entomopathogenic nematodes (*Steinernema* spp.) against key insect pests of Gujarat. M. Sc. thesis submitted to Gujarat Agricultural University, S. K. Nager 385 506, pp. 64

Patel, N.B., Vyas, R.V. and Patel, D.J. (1996). Compatibility of nematophagous fungus *Paecilomyces lilacinus* with key nematicides. *In*: Proc. of National symposium Towards Green and Sustainable Pesticides. December 4-6, 1996, Society of Pesticide Science, IARI, New Delhi, India. pp 6 (Abrs.).

Patel, N.B., Vyas, R.V. and Patel, D.J. (1998). Efficacy of biocontrol fungi *Paecilomyces lilacinus* for management of *Meloidogyne javanica* pt.2 on groundnut. *In*: Proc. of the National Symposium on Rational Approach in Nematode Management for Sustainable Agriculture (NSI), Nov. 23-25, 1998, Gujarat Agricultural University, Anand, India. pp.23-24.

Patel, N.S. and Vyas, R.V. (2001). Evaluation of native entomopathogenic nematodes (*Steinernema* spp.) against soil insect pests. Paper presented in National Congress on Centenary of Nematology in India – Appraisal and Future Plans, jointly sponsored b Nematological Society of India, Indian Council of Agricultural Research, National Academy of Agricultural Sciences, Department of Science and Technology and Society of Plant Protection Sciences at Division of Nematology, IARI, New Delhi-110 012 on Dec. 5-7,2001. P.113.

Patel, N.S., Vyas, R.V. and Patel, D.J. (2001). Biological suppression of cutworm , Agrotis ipsilon using entomopathogenic nematode (*S. glaseri*) on potato in pots. Journal of Mycology and Plant Pathology, 31(1): 120 (Abst.).

Patel, R. C.; Singh, R. and Patel, P. B. (1968). Nuclear Polyhedrosis of the Gram pod borer, *Heloithis armigera* J. Econ. Entomol., 61 (1) : 191-193.

Patel, R.R.; Patel, D.J.and Patel, B.A. (1991). Influence of organic amendments on growth and sporulation of nematophagous fungus, *Paecilomyces lilacinus* (Thom.) Samson. Curr. Nematol. 2(1): 39-40.

Patel, R.R.; Patel, D.J. and Patel B.A. (1992). Growth and sporulation of nematophagous fungus, *Paecilomyces lilacinus* (Thom.) Samson as influenced by different cereal media. Curr. Nematol. 3(1): 99-100.

Pawar, V. M.; Chandurawar, R. D.; Kadam, B. S.; Thombre, U. T.; Dhawandkar, S. D. and Seeras,

Pharindra (2000). Comparative bioefficacy of Eco friendly. Microbial bioagents against *Heliothis* (*Helicoverpa armigera*) (Huber) on chickpea (*Cicer arietinum* L.) M.Sc. Thesis submitted to Gujarat Agricultural University S.K. Nagar. P. 55.

Prasad, R.D. and Rangeshwaran, R.(2000). J. Mycol. Pl. Patho., 30(2) :216-220

Prasad, S. S. V. and Tilak, K. V. B. R. (1972). Aerobic spore-forming bacteria from a root-knot nematode infested soil. Indian J. Micro. 11 : 59-60.

Prasad, S. S. V., Tilak, K. V. B. R. and Gollakota, K. G. (1972). Role of *Bacillus thuringiensis* var. *thuringiensis* on the larval survivability and egg hatching on *Meloidogyne* spp. the causative agent of root-knot disease. J. Invt. Pathol. 20 : 377.

Prasad, V.D.; Jayaraj, S.; Rabindra, R.J. (1990). Susceptibility of gram caterpillar, *Heliothis armigera* Hbn (Noetudiae: Lepidoptera) to certain Entomogenous fungi. J. Biol. Cont., 4(1): 44-47.

Rabindra, R. J.; Jayaraj, S. and Balasubramanian, M. (1989). Efficacy of nuclear polyhydrous virus for the control of *Heliothis armigera* (Hubner) infesting sunflower. J. Ent. Res., 9 (2): 246-248.

Rabindra, R. J.; Ethiraju, S. and Jayaraj S. (1991). Efficacy of Nuclear polyhedrosis virus formulations against *Heliothis armigera* (Hub.) sunflower J. Biol. Cont., 4 (2) : 130-131.

Rabindra, R. J.; Sathiah, N. and Jayaraj, S. (1992). Efficacy of nuclear polyhedrosis virus against *Heliocoverpa armigera* (Hub.) on *Helicoverpa* resistant and susceptible varieties of chickpea. Crop Prot. 11 (4) : 320-322.

Rabindra, R. J.; Muthuswami, M. and Jayaraj S. (1994). Influence of host plant surface environment on the virulence of nuclear polyhedrosis virus against *H. armigera* (Hub.) (Lep., Noctuidae) larvae. J. appl. Entomol., 118 (4-5) : 453-460.

Rao, M. S., Gower, S.R., Bourne, J.M. and Parvatha Reddy, P. (1997). Management of *Meloidogyne incognita* in tomato nurseries by integration of *Golomus deserticola* with *Verticillium chlamydosporium*. J. Plant Dis. Protec. 104 : 419-422.

Rao, M. S., Parvatha Reddy, P. and Nagesh, M. (1997). Management of root-knot nematode *Meloidogyne incognita* on tomato by integration of *Trichoderma harzanium* with neem cake. J. Plant Dis. Protc. 104 : 423-425.

Saikia, M.K. and Roy, A.K. (1994). Efficacy of *Paecilomyces lilacinus* on the reduction of attack of *Meloidogyne incognita* on okra. Indian J. Nematol. 24(2): 163-167.

Sankaranarayanan, C. and Rajeswari Sundarababu. (1998). Effect of oil cakesand nematicides on growth of black-gram (*Vigna mungo*) inoculated with VAM fungus (*Glomos fasciculatum*) and root knot nematode (*Meloidogyne incognita*). Indian J. Nematol. 27(1): 128-129.

Sankaranarayanan, C. and Rajeswari Sundarababu. (1999). Effect of phosphatic fertilizers on the interaction of *Glomos fasciculatum* and *Meloidogyne incognita* on black-gram (*Vigna mungo*). Indian J. Nematol. 29(1): 44-47.

Sankaranarayanan, C., Hussaini S.S., Sreerama Kumar, P. and Rangeshwaran, R. (2000). Granular application of antagonistic fungi for the biological control of *Meloidogyne incognita* on tomato. Indian J Nematol. 30(2) : 157-161.

Sankaranarayanan, C., Hussaini, S. S., Kumar, B. S. and Prasad, R. D. (1997). Nematicidal effect of fungal filtrates against root-knot nematodes. J. Biol. Cont. 11 (1-2): 37-41.

Santharam, G.; Balasubramanian, M. and Chelliah, S. (1981). Control of *Heliothis armigera* (Hub.) on red gram (*Cajanus cajan* L.) with a nuclear polyhydrosis virus and insecticides. Madras Agric. J. 68: 417-420.

Santiago, D.R. (1991). Control of cabbage insect pests by *Metarhizium anisopiae* (Metschnikoff) (Philippines). College Laguna (Philippines). p.90.

Saxena, H and Ahmad, R. (1999). Field evaluation of *Beauveria bassiana* (Belsamo) vuikemin against *Helocoverpa armigera* (Hubner) infecting chickpea. J. Biol. Cont. 17: 93 – 96.

Shanti, A. and Sivakumar, C. V. (1995). Biological potential of *Pseudomonas flourescens* (Migula) against root-knot nematode, *Meloidogyne incognita* on tomato. J. Biol. Cont. 9: 113-115.

Sharma, A. and Trivedi, P.C. (1987). Screening of substrate suitable for the growth of *Paecilomyces lilacinus*. Int. Nematol. Network Newsl. 4(4): 24 - 25.

Sharma, B. K. and Bansandrai, A.K. (2000). J. Mycol. Pl. Patho. 30(1) :16-18

Shelton, A.M.; Vanderberg, J.D.; Ramos, M.; and Wilsey, W.T. (1998). Efficacy and persistence of *Beauveria bassiana* and other fungi for control of diamondback moth (Lepidoptera:Plutellidae) on cabbage seedlings. J. Entom. Sci. 33(2): 142-151.

Shenhmar, M.; Sehhan, B. S. and Brar, K. S. (1996). Efficacy of some bio-pesticide for the control of *Helicoverpa armigera* Hubner. Indian J. Pl. Prot., 24 (1 & 2) : 132-133.

Sivakumar, C. V. (1998). Bacterial antagonists for suppression of plant parasitic nematodes. In : S. P. Singh and S. S. Hussaini (Eds.) Biological Suppression of Plant disease, Phytoparasitic nematodes and Weeds. Project Directorate of Biological Control, Bellary Road, Bangalore 560 024, India : pp 128-137.

Sivakumar, C.V., Bhaskaramanian, M.D. and Balasaraswathy, S. (1993). Control of root-knot nematode, *Meloidogyne araneria* in brinjal nursery with fungus , *Paecilomyces lilacinus* and systemic nematicide carbofuran. J. Biol.Cont. 7(1) : 49-50.

Somasekhar, N. and Gill, J.S. (1990). Penetration and multiplication of *Meloidogyne incognita* as influenced by *Pasteuria penetrans*. Indian J. Nematol. 20(2): 213-218.

Somasekhar, N. and Gill, J.S. (1991). Efficacy of *Pasteuria penetrans* alone and in combination with carbofuran in controlling *Meloidogyne incognita*. Indian J. Nematol. 21(1):61-65.

Sreeja, T. P., Eapen, S. J. and Ramana, K. V. (1996). Occurrence of *Verticillium chlamydosporium* Goddard in black pepper *Piper nigrum* L. garden in Kerala. Indian J. Spices and Aromatic Cr. 5 : 143-147.

Sundararaja, P., Sudha, S. and Rohini, Iyer. (1990). Reaction of Subabul varieties to root-knot nematode and V A mycorrhizae on Subabul seedlings. Indian J. Nematol., 20 : 43-48.

Swarup, G. and Dasgupta, D.R. (1986). Plant Parasitic Nematodes of India Problems and Prospects, Director, IARI (Allied Publisher), New Delhi,India. 497pp.

Therumala Rao, S.K. and Sitaramaiah(2000). J. Mycol. Pl. Patho.,30 (2):221-224

Thomas, G. V., Sundararaja, P. and Ghai, S. K. (1989). Individual and interactive effect of V A mycorrhizal fungi and root-not nematode *Meloidogyne incognita* on Cardamom. Trop. Agric. 66 : 21-24.

Umarao, Sudershan Ganguly, R.V.Vyas and A. K. Ganguly (2002). Molecular characterization of *Steinernema thermophilum* Ganguly and Singh (Nematoda: Steinernematidae). International Journal of Nematology 12 (2): pp. 215-219

Ushamalini, C., Rajappan, K. and Gangadharan, K. (1997). Indian Phytopath., 50 (4) : 504-507

Vandenberg, J.D.; Shelton, A.M.; Wilsey, W.T. and Rama, M. (1991). Assessment of *Beauveria bassiana* sprays for control of diamond backmoth (Lepidoptera : plutillidae) on crusifersh. J. Econ. Entomol. 91 (3) : 624 – 630.

Veen, K.H. and Ferron, P. (1986). A selective media for the isolation of *Metarhizium anisopliae*. J. Insect Pathol. 8: 268-69.

Vyas, H. G.; Lakhohaura, B. D. (1996). Evaluation of *Heliothis* nuclear polyhedrosis virus for control of *Heliothis armigera* on chickpea at Pantnagar, (U.P.) GAU Res. J. 21 (2) : 50-54. 29: 795-97.

Vyas, R.V., Yadav, D.N. and Patel R.J. (1990). Studies on the efficacy of *Beauveria brongniartii* (Sacc). Petch against white grub. Ann. Biol. 6(2): 123-28.

Vyas,R.V., Yadav, D.N. and Patel R.J. (1990). A note on compatibility of *Beauveria brongniartii* (Sacc).Petch with some pesticides used in groundnut pest management. Ann. Biol., 6(1):21-26.

Vyas, R.V., Yadav, D.N. and Patel R.J. (1991). Mass production of a entomogenous fungi *Beauveria brongniartii* on solid substrates. Indian J. Exp. Biol. 29: 795-97

Vyas, R.V. and Yadav, D.N. (1992). Infectivity of entomopathogenic nematode to the two soil dwelling lepidopteron. Annals of Biology, 8(1) : 59-63.

Vyas, R.V., Jani, J.J. and Yadav, D.N. (1992). Effect of some natural pesticides on entomogenous muscardian fungi. Indian J. Exp. Biol., 30:435-36.

Vyas, R.V. and Yadav, D.N. (1992). Possibilities of using entomophilic nematode for the management of three major pests of agricultural crops. Journal of Nuclear Agriculture and Biology, 21(3) :268-69 .

Vyas, R.V. and Yadav, D.N. (1993). *Steinernema glaseri* (Steiner) Travassos, a biological control agent of root grub *Holotrichia consangainea* Blanchard. Pakistan Journal of Nematology, 11 (1) : 41-44.

Vyas, R.V., Patel, H.R., Patel, B.A. and Patel, D.J. (1995). Parasitism of muscardine fungi on *Meloidgyne incognita* eggs. Curr. Sci. 69(8) : 634-635.

Vyas, R.V., Patel, B.A., Patel, D.J. and Patel, R.S. (1996). Integrated management of root-knot nematodes in chickpea. Pakistan.J.Nematol. 14 : 117-119.

Vyas, R. V., N.R. Joshi, N.B. Patel and D.J. Patel (1997). Bioassay of entomopathogenic nematode, *Steinernema glaseri* on *Spodoptera litura* larvae with and with out adjuvant. Paper presented in Gujarat Science Congress organized by Gujarat Academy of Sciences, Ahmedabad at V.P. Science College, Sardar Patel University, Vallabh Vidyanagar-388 120 on Feb. 15-16, 1997.P.6.

Vyas, R.V., Patel, H.R., Patel, D.J., Patel, N.B. and Joshi, N.R. (1997). Biological suppression of root-knot nematode-white grub pest complex attacking groundnut. *Int.* Arachis News Lett. 17 :40-41

Vyas,R.V., Patel,N.B. and Patel, D.J. (1998). Integrated management of root-knot nematodes in cotton . *In*: Proc. of the National Symposium on Rational Approach in Nematode Management for Sustainable Agriculture (NSI), Nov. 23-25, 1998, Gujarat Agricultural University, Anand, India. pp 51-52.

Vyas , R.V., Pharindera Yadav, Choudhary, R. and Patel, N.S. (1998). Natural incidence of Entomopathogenic nematodes in Central - North Gujarat. *In*: Proceedings of the National Symposium on Rational Approach in Nematode Management for Sustainable Agriculture, held on Nov. 23-25, 1998 at Gujarat Agricultural University, Anand. pp 86-88.

Vyas, R.V., Patel, N.S. and Patel, D.J. (1999). Mass production technology for entomopathogenic nematode, *Steinernema* sp. Indian Journal of Nematology, 29:178-181.

Vyas, R. V. (2000) Entomopathogenic Nematodes - A New Tool for Management of Insect Pests. Paper presented as participants seminar at International Centre for Genetic Engineering and Biotechnology , New Delhi during practical course on Insect Resistance : Role of BT toxin held on December 11-22,2000.

Vyas,R.V. , Patel,N.B. and Patel, D.J. (2000). Biological control of Root-knot nematode and white grub pest complex on groundnut in field. J.Mycol.Pl.Pathol. , 30 (2) : 258 (Abrs.).Vyas, R.V., Patel, N.B., Patel, H.R. and Patel, D.J. (2001). Search for a cheaper substrate for mass production of nematophagous fungi *In vitro*. J. Mycol. Pl. Pathol. 31 : 106-107(Abrs.).

Vyas, R.V., Patel, B.A., Patel, D.J. and Patel, R.S. (2000). Integration of effective technologies for economic management of root-knot nematodes in tomato nursery. *In*: Proc. of Indian phyto pathological society Golden jubilee International Conference on Integrated plant disease management for sustainable agriculture Vol.2 , Nov. 10-15, 1997 , IARI, New Delhi, India.pp 697-698.

Vyas, R.V.; Pharindra Yadav; Ghelani, Y. H.; Chaudhary, R. K.; Patel, N. B. and Patel, D. J. (2001a). *In vitro* mass production of native *Steinernema* sp. Annals of Plant Protection Sciences, 9: 77-78.

Vyas, R. V., Ghelani, Y. H., Patel, N. B. and Patel, D.J. (2001b). Bioefficacy of Entomopathogenic Nematode against *Helicoverpa armigera* on Pulse Crops. Paper presented in National Congress on Centenary of Nematology in India – Appraisal and Future Plans, jointly sponsored by Nematological Society of India, Indian Council of Agricultural Research, National Academy of Agricultural Sciences, Department of Science and Technology and Society of Plant Protection Sciences at Division of Nematology, IARI, New Delhi-110 012 on Dec. 5-7,2001. P: 114

Vyas, R. V., Patel, N.B., Parul D. Patel and Patel, D.J. (2001c). Field Efficacy of Entomopathogenic Nematode (*Steinernema* nr. *riobrave*) against Cotton Bollworms. Paper presented in National Congress on Centenary of Nematology in India – Appraisal and Future Plans, jointly sponsored by Nematological Society of India, Indian Council of Agricultural Research, National Academy of Agricultural Sciences, Department of Science and

Technology and Society of Plant Protection Sciences at Division of Nematology, IARI, New Delhi-110 012 on Dec. 5-7,2001. P: 115.

Vyas, R. V. (2002). Entomopathogenic nematodes – A new tool for management of insect pests : REVIEW OF EPN WORK *In*: Current Status of Research On Entomopathogenic Nematodes in India. (Eds) S. S. Hussaini, R. J. Rabindra and M. Nagesh), Project Directorate of Biological Control (ICAR), Bellary Road, Bangalore-560 024: pp. 113-120

Vyas, R. V., Patel, N. B., Parul Patel and Patel, D. J. (2002a). Efficacy of entomopathogenic nematode against *Helicoverpa armigera* on pigeonpea. International Chickpea & Pigeonpea News Letter, 9: 43-44.

Vyas R. V.; Maghodia A. B.; Patel, Biren A. and Patel, D. J. (2003) Interaction of root-knot nematode, *Meloidogyne incognita* and entomopathogenic nematode, *Steinernema riobrave* on okra J. Mycol. Pl. Pathol., 33 (3) : 485

Vyas, R. V., Maghodia, A. B., Patel, Biren A. and Patel, D. J. (2004). Interaction between root–knot nematodes and *Steinernema riobrave* on okra (Abelmoschus esculentus L.). International J Nematol. 14 : 186-190.

Vyas, R. V., Maghodia A. B.; Patel B. A. and D. J. Patel. (2005a). *In Vitro* Testing of *Xenorhabdus* metabolites against Groundnut Collar-Rot Fungus, *Aspergillus niger*. International Arachis Newsletter, 25: 34-36.

Vyas, R. V., Maghodia A. B., Patel Biren A. and Patel, D. J. (2005b). Bioefficacy of native *Steinernema riobrave* against *Meloidogyne javanica* pathotype 2 and *Holotrichia consanguinea* complex on peanut (*Arachis hypogaea* L.). International J Nematol. 15: 207-209.

Vyas, R. V., A. B. Maghodia, Biren A. Patel and D. J. Patel (2005c). Field Efficacy of entomopathogenic nematode,*Stinernema riobrave* and their exotoxins against root knot nematodes, *Meloidogyne spp*. on okra (Indian Journal of Nematology special issue, Communicated).

Walia, K. and Swarup, G. (1985). Effect of some fungi on nematode hatching and larval root penetration. Indian J. Nematol. 15(2) : 174-176.

Walia, K. and Swarup, G. (1985). Isolation of nematophagous and parasitic fungi from soil and other sources in India. Indian J. Nematol. 15(2): 253.

Zaki, F.A. and Bhatti, D.S. (1990). Efficacy of castor (*Ricinus communis*) and the biocontrol fungus, *Paecilomyces lilacinus* on *Melidogyne incognita*. Nematologica 36(1):114 -122.

—•—

Annexure I
(Composition of Culture Media)

(A) Media for the Biofertilizers

Rhizobium

1. Yeast Extract Mannitol Congored Agar Medium

Mannitol	10 g
K_2HPO_4	0.5 g
Yeast Extract	1 g
$MgSO_4$	0.2 g
NaCl	0.1 g
$CaCO_3$	3 g
Congored (1: 400 i.e. 250 mg/100 ml)	10 ml
Agar	20 g
pH	7.0
Distilled water	1000 ml

Medium for Quality Control of *Rhizobium*

2. Glucose Peptone Agar

Glucose	5 g
Peptone	10 g
NaCl	5 g
Agar	20 g
Distilled water	1000 ml
pH	7.2

3. Hofer's Alkaline Broth

Mannitol	10 g
K_2HPO_4	0.5 g
$MgSO_4$	0.2 g
NaCl	0.1 g
Yeast Extract	0.5 g
$CaCO_3$	0.05 g
Distilled water	1000 ml
pH (Adjusted with 2 N NaOH)	10-11
Bromothymol blue solution	1 ml

4. Growth on Lactose Agar

Lactose	10 g
K_2HPO_4	0.5 g
Yeast extract	0.5 g
$MgSO_4$	0.2 g
NaCl	0.1 g
$CaCO_3$	3 g
Distilled water	1000 ml
pH	7.0
Agar	15 g

5. Testing nodulating ability of Rhizobium (Jenson's Plant Nutrient)

Calcium hydrogen phosphate	1 g
Potassium monohydrogen phosphate	0.2 g
Magnessium sulphate	0.2 g
Sodium chloride	0.2 g
Iron chloride	0.1 g
Agar-agar	8 g
Distilled water	1000 ml
pH	6.8

Azotobacter

1. *Waksman No. 77 Medium*

Mannitol	10 g
K_2HPO_4	0.5 g
$MgSO_4$	0.2 g
NaCl	Trace
$FeCl_3$	Trace
Distilled water	1000 ml
pH	7.0
Agar	15 g

2. *Ashby's Mannitol Agar*

Mannitol	20 g
K_2HPO_4	0.2 g
$MgSO_4.7H_2O$	0.2 g
NaCl	0.2 g
K_2SO_4	0.1 g
$CaCO_3$	5 g
Agar	15 g
Distilled water	1000 ml

3. *Jensen's Medium*

Sucrose	20 g
K_2HPO_4	1 g
$MgSO_4.7H_2O$	0.5 g
NaCl	0.5 g
$FeSO_4$	0.1 g
Na_2MoO_4	0.005 g
$CaCO_3$	2 g
Agar	15 g
Distilled water	1000 ml

Azospirillum

1. *Dobereiner's nitrogen free malic acid semi - solid medium*

Malic acid	5 g
KOH	4 g

K_2HPO_4	0.5 g
$MnSO_4$	0.01 g
$FeSO_4.7H2O$	0.05 g
$MgSO_4$	0.01 g
NaCl	0.02 g
Na_2MoO_4	0.002 g
Distilled water	1000 ml
Bromothymol blue (0.5% alc.soln)	2 ml
Agar	1.75 g
pH	6.6-7.0

(For mass production add 1.00 g Ammonium chloride per liter of the medium to enhance biomass production at a faster rate)

2. *Rojo Congo Medium*

DL malic acid	5 g
K_2HPO_4	0.5 g
$Mg\,SO_4$	0.2 g
NaCl	0.1 g
Yeast extract	0.5 g
$Fe\,Cl_3.\,6H_2O$	0.015 g
KOH	4.8 g
Congo Red (1: 400 aq. Solution)*	15 ml
Agar	18 g
Distilled water	1000 ml
pH (with 0.1N KOH)	7.0

* Sterilized separately and added to the medium after autoclaving.

3. *Potato Infusion Agar medium*

Potato	200 g
Malic acid	2.50 g
KOH	2 g
Sucrose	2.50 g
Vitamin solution	1 ml
Agar	15 g
Distilled water	1000 ml

Phosphorus solubilizing bacteria

1. Soil Extract Medium

Soil extract	100 ml
Glucose	1 g
Agar	2 g

Medium is sterilized in 100 ml lots and after sterilization 5 ml of 10% K_2HPO_4 and 10 ml of 10% $CaCl_2$ should be added. pH is adjusted to 7.0 with sterile 0.1 N NaOH.

2. Pikovskaya's Medium

Glucose	10 g
Tri calcium phosphate	5 g
$(NH_4)_2SO_4$	0.5 g
KCl	0.2 g
$MgSO_4.7H_2O$	0.1 g
$MnSO_4$	Trace
$FeSO_4$	Trace
Yeast extract	0.5 g
Agar	15 g
Distilled water	1000 ml

3. Nutrient Glucose Broth

Beef extract	3 g
Peptone	5 g
Glucose	5 g
NaCl	5 g
Distilled water	1000 ml
pH	6.5-7.0

Acetobacter diazotrophicus

1. Diluted Cane Juice Medium

Semisolid LGIP medium	250 ml
Sugarcane juice	250 ml
Distilled water	500 ml

2. *Semisolid LGIP Medium*

K_2HPO_4	0.2 g
KH_2PO_4	0.6 g
$MgSO_4$	0.2 g
$CaCl_3$	0.02 g
Na_2MoO_4	0.002 g
$FeCl_3$	0.01 g
Bormothymol (0.5 % in 0.2 N KOH)	5 ml
Cane sugar	100 g
Agar	1.8 g
Distilled water	1000 ml
pH	6.0

3. *Semisolid acetic LGIP Medium*

Semi solid LGIP medium was acidified with acetic acid to pH 4.5 and agar concentration was increased to 2.2 g l^{-1}.

(B) Media for Microbial Pesticides

Fungi

1. *Potato Dextrose agar (PDA)*

Potatoes infusion	200 g
Dextrose	20 g
Agar	15 g
Distilled Water (D/W)	1000 ml
pH	4.5

2. *Selective Media for Paecilomyces lilacinus*

Glucose	10 g
Peoptone	10 g
Agar-agar	35 g
D/W	1000 ml
Antibiotics supplement	
N-dodicylguanidine mono acetate)	10 µg/ml
Chloremphenicol	500 µg/ml
pH	4

3. Composition of Selective Medium for T. viride

Magnesium sulphate	0.20 g
Dipotassium Hydrogen Phosphate	0.90 g
Ammonium Nitrate	1 g
Potassium Chloride	0.15 g
Glucose	3 g
Metalaxyl	0.30 g
Rose Bengal	0.15 g
Chloramphenicol	0.25 g
Agar	15 g
Distilled water	1000 ml
pH	6.00 - 6.50

4. Molasses Yeast Medium for Fermentation Process

Molasses (Note: Instead of Molasses, can use Jaggery 20 g)	30 g
Yeast Extract	5 g
Distilled water	1000 ml
pH	6.00 – 6.5

B. thuringiensis

(1) Yeast Extract Mannitol Polymixin-B Agar (YEMA)

Part A

Peptone	10 g
Meat Extract	1 g
D-Mannitol	10 g
Sodium chloride	10 g
Phenol red	0.025 g
Agar- agar	15 g
D/W	1000 ml

Part B

Polymixin-B Sulphate 100 units/100 ml	
Egg yolk emulsion (10%) 100 ml/lit.	
D/W	1000 ml
pH at 25 °C	7.1 ± 0.2

(2) Nutrient Agar

Peptone	5 g
Sodium chloride	5 g
Beef Extract	1.5 g
Yeast Extract	1.5 g
Agar - agar	15 g
pH at 25° C	7.4 $\pm$ 0.2
D/W	1000 ml

(3) Luria Broth (LB)

Tryptone	10 g
Sodium chloride	5 g
Yeast extract	5 g
pH at 25° C	7.4 $\pm$ 0.2
D/W	1000 ml

(4) Liquid Media Ingredients

Soybean meal	15 g
Dextrose	5g
Corn starch	5g
$Mg SO_4 . 7 H_2O$	0.3 g
$Fe SO_4 . 7 H_2O$	0.02 g
$Zn SO_4 . 7 H_2O$	0.02 g
$Ca CO_3$	1 g
D/w	1000 ml

(5) Semi Solid Media Ingredients

Wheat bran	545 g
Expanded per liter	380 g
Soybean meal	62 g
Dextrose	36 g
Lime	3.6 g
NaCl	0.9 g
$CaCl_2$	0.29 g
Pure Water	160 ml

EPN

1. Dog Food Media

Pulverized dry dog food	100 g
1% agar	500 ml

2. Egg Yolk Yeast Agar Medium

Hen Egg Yolk	10 g
Yeast Extract	1 g
Agar Agar	2 g
D/W	1000 ml

3. Liquid Medium

Chicken extract (10%)	10 ml
Yeast extract	0.5 g
Beef extract	0.5 g
Distilled water	90 ml
pH	6.8 -7.0

Pseudomonas fluorescens

1. Selective Medium for Pseudomonas

Peptone	20 g
K_2HPO_4	1.5 g
$MgSO_4$	1.5 g
Glycerol	10 ml
Agar	20 g
Ampicillin	50 mg
Cycloheximide	12.5 mg
Chloramphenicol	12.5 mg
D/W	1000 ml

2. King' B Agar Medium

Peptone	20 g
$MgSO_4$	1.5 g
KH_2PO_4	1.5 g
Glycerol	10 ml
Distilled water (D/W)	1000 ml
pH	6.8 to 7.2
Agar	20 g

Annexure II
(Finance for Biofertilizer Unit and Commercial Biofertilizer Producers in India)

FINANCE FOR BIO-FERTILIZERS PRODUCTION UNIT (BPU)

(A) Application for getting Finance from NABARD

To be completed by the Executive/Officer of the bank forwarding the scheme

Note: Tick () in the boxes to signify that the relevant information has been furnished in the scheme.

1) GENERAL

a) Project background, specification of the scheme area and arrangements for acquisition of land.

b) Nature and objective of proposed development and whether as an independent item or an integral part of the project.

c) Name (s) of the financing banks(s)/branch (es)

d) Status of implementing agency and Promoter's Bio-Data

2) TECHNICAL ASPECTS

a) Source of technology

b) Type of bio-fertilizer produced (single/multi product)

c) Arrangements for procurement of microbial strains, storage and multiplication.

d) Method of production (Shake Flask/Fermentor)

e) Provision for sterilization, if so particulars thereof

f) Carriers used (Lignite/Charcoal/Vermiculite/Peat)

g) Availability of skilled labour for handling of stock.

h) Market survey and assessment of the demand in the area - potential

i) Detailed cost estimates of the unit and auxiliary structures prepared as per approved schedule of rates applied to site.

j) Details of drawings of structures including layout plans of all civil works.

k) Phasing of the programme with financial outlay

l) Comments on technical feasibility of the project

m) Arrangements for quality control

n) Arrangements for marketing

3) FINANCIAL ASPECTS

a) *Lending terms:* Rate of interest, grace period, repayment period, down payment, nature of security, availability of government guarantee for bank loan/refinance (wherever applicable), source and extent of availability of subsidy/grant-in-aid, if any, etc.

b) Year-wise physical, financial programme, bank loan and refinance requirement.

c) Estimates of unit-wise aggregate income, expenditure and surplus. Comments on financial viability of the project along with cash flow, B/C ratio, net present worth, financial rate of return, etc.

d) Whether any marketing arrangements with State Govt/Board/ Corporation, if any.

e) Whether any support price/marketing policy exist in the state, if so particulars thereof.

f) Whether the production technology complies with quality control (BIS/ISI).

g) Whether subsidy is available to the buyers

h) Comments on the financial position of the borrowers/ implementing agency. In case of companies, partnership firm, etc., an analysis of their financial position, debt-equity and profitability along with copies of audited financial statements, for last three years.

Signature and Designation of
Bank Officer

(B) Application for Getting Subsidy from Government

Title	National Project on Development and Use of Biofertilisers
Types	**Central Sector Scheme**
Objectives	i) Production and distribution of Biofertilisers (BFs) ii) Develop Standards for different BFs and Quality control iii) Release of grants for setting up BF units iv) Training and Publicity
Salient features	i) Under this scheme the Government provides non-recurring grants-in-aid up to Rs.20.00 lakhs for setting up biofertiliser production units of 150 MT capacity. The grant-in-aid is offered to State Departments of Agriculture/cooperatives/public sector undertakings of fertilizers; NGOs and private agencies provided their proposals are received from respective State Governments. ii) One time grant-in-aid to the extent of Rs.1.50 lakhs is provided for establishment of Blue-green algae sub-centre to produce 30-40 tonnes of BGA/annum provided their proposals are received from respective State Governments. iii) This scheme is being implemented in the country with the help of a National and 6 Regional centers for training, field demonstration and its promotion and also through Department of Agriculture & Cooperation i.e. release of one time grant-in-aid for setting up BF units.
Pattern of Assistance	Financial assistance of 20 lakhs is being provided for creating a production capacity of 150 tons of biofertilisers per annum and Rs.10 lakhs for 75 tons capacity per

	annum. The grants are released through the State Governments in two installments.
Eligibility	The grants may be released to State Departments of Agriculture/cooperatives/public sector undertakings of fertilizers, NGOs and also to private entrepreneurs provided their proposals are received through respective State Govt.
Procedures to apply	Proposed to implement the scheme through NABARD/ NCDC during 2001-2002. Revised guidelines are under preparation.
Person to be contacted	The Joint Secretary Fertilizer Division, Department Agriculture and Cooperation Krishi Bhawan, New Delhi
Date of Start/ duration	The scheme was launched in March, 1983 and approved to continue during IX Plan with an estimated expenditure of Rs.19.30 crores.
Implementation	i) National Biofertiliser Development Centre has been
Status	established at Ghaziabad with six Regional Biofertiliser Development Centres each at Jabalpur, Hissar, Nagpur, Bangalore, Bhubaneswar and Imphal.
	ii) Under this scheme 74 BF production units have been established with central financial assistance having annual production capacity of 8475 tonnes.
Additional information	Biofertilisers are being considered important component of Integrated Plant Nutrient System (IPNS) and Organic farming for sustaining the agriculture.

Commercial Biofertilizers in India

S. No.	Producer & Address	Biofertilizer
01	A. P. Agriculture University Bacterial Scheme Agricultural Research Station, Amaravathi – 522 020 Guntur, A.P.	*Rhizobium*
02	A.P. Department of Agricultural, Bacterial Culture Production Lab. Regional Soil Testing Laboratory Rajendranagar, Hyderabad - 500 030	*Rhizobium*

03	Ajay Bio Tech Lab. Ltd. 17A/19, Kubera Park, Kondhwa Naka Pune - 411 040, Maharastra	*Rhizobium*, PSM
04	Assam Agriculture University Jorhat - 785 013, Assam	***Rhizobium***
05	Assam Agro Indistries Corp. Agricultural Chemistry Research Station Guwahati, Assam	*Rhizobium*
06	Azo Fertilizers Pvt. Ltd. 1063/89, I Main, 7th Cross, Vidyaranya Puram Mysore - 570 008, Karnataka	*Rhizobium* *Azotobacter* *Azospirillum*
07	BAIF Development Research Research Foundation Opp. Pune-Nagar Road, Wagholi Pune - 412 207, Maharastra	*Rhizobium* VAM
08	Bidhan Chandra Krishi Vishwa Vidyalaya, Nodule Research Laboratory, Mohanpur-741 235, Dt. Nadia, West Bengal	*Rhizobium*
09	Birsa Agricultural University Ranchi - 834 006, Bihar	*Rhizobium*
10	Farm Organics India Pvt. Ltd. 1126/1, Shri Ram Complex, Opp. Jagtap Dairy, Model Colony, Pune - 411 016	
11	Godavari Fertilizer & Chemicals Ltd. Banjara Hills, 8-2-696, Road No.12, Hyderabad - 500 016	***Rhizobium***
12	National Bio fertilizer Development Centre, 204-B Wing, CGO Complex-II, Kamala Nehru Nagar, Ghaziabad - 201 002, U.P.	*Rhizobium* *Azotobacter* *Azospirillum* BGA, PSM
13	Government of India Regional Bio Fertilizer Development Centre, VCA Complex, Civil Lines, Nagpur - 440 001, Maharastra	*Rhizobium* *Azotobacter* *Azospirillum* BGA, PSM

14	Government of India Regional Bio Fertilizer Development Center, 213, Ravindra Nagar, Adhatlal Jabalpur – 482 004, M.P.	*Rhizobium Azotobacter* *Azospirillum* BGA, PSM
15	Government of India Rigional Bio Fertilizer Development Centre, A – 156 Shahid Nagar, Bhubaneswar – 751 007, Orissa	*Rhizobium* *Azotobacter*
16	Government of India Regional Bio Fertilizer Development Centre, 34-II Main Road, Near Baptist Hospital, Hebbel, Bangalore – 560 024, Karnataka	*Rhizobium* *Azotobacter* *Azospirillum* BGA, PSM
17	Government of India Regional Bio Fertilizer Development Centre, 18, Sirsa Road, Near Sewak Sabha Hospital, Hissar – 125 001, Haryana	*Rhizobium* *Azotobacter* *Azospirillum* BGA, PSM
18	Government of India Regional Bio Fertilizer Development Centre, Agriculture College Campus, Irooisemba, Imphal – 759 001, Manipur	*Rhizobium* *Azotobacter* *Azospirillum* BGA, PSM
19	Gujarat State Co-Operative Marketing Federation, Sahakar Bhawan, Relief Road, Ahmedabad – 380 001, Gujarat	***Rhizobium***
20	Gujarat State Fertilizer Co. Ltd. P.O. Fertilizer Nagar, Vadodara – 391 750, Gujarat	*Rhizobium* *Azotobacter Azospirillum* *Acetobacter*, PSM
21	Hans Biogenics, 14-146, Gandhipuram 2, Rajamundry – 533 103, A.P.	*Rhizobium*
22	Haryana Agricultural University Dept. of Micro Biology, Hissar – 125 001, Haryana	*Rhizobium* *Azotobacter* *Azospirillum*
23	H. P. Department of Agriculture Central Laboratory, Hawthrom Villa, Shimla – 171 004, H. P.	*Rhizobium*

24	Indian Agricultural Research Institute, Division of Micro Biology, New Delhi - 110 012	*Rhizobium*, BGA *Azotobacter*, PSM, *Azospirillum*, VAM
25	Indian Farmers Fertilizer Co-Op. Ltd. 34, Nehru Place, New Delhi	*Rhizobium*, VAM *Azotobacter*, PSM *Azospirillum*, BGA
26	Indian Organic Chemicals, 655, Sukrawar Peth, Gadikhana Shivaji Road, Pune - 411 002	*Rhizobium* *Azotobacter* PSM
27	K-Ferts Ltd., W-4 MIDC Ind. Area, Nanded - 431 603, Maharastra	***Rhizobium***
28	Karnataka Dept. of Agriculture Rhizobium Culture Production Lab. Hebbal, Banglore - 560 024	*Rhizobium* *Azospirillum* *Azotobacter*, PSM
29	Karnataka Dept. of Agriculture Soil Health Centre, Gulbarga, Karnataka	*Rhizobium* *Azospirillum* *Azotobacter*, PSM
30	Kerala Dept. of Agriculture, Vikas Bhawan, Trivandrum - 659 035, Kerala	*Rhizobium* *Azospirillum* *Azotobacter*, PSM
31	Krishak Bharati Co-Operative Ltd. Plot No. 10A, Sector-II, Noida Dt. Ghaziabad, U.P. - 201 301	*Rhizobium* *Azospirillum* *Azotobacter*, PSM
32	Krishi Mitra Bioo Products I Pvt. Ltd. 12, Ganeshwadi, Prim, K.R. Kanitkar Peth, Opp. F. C. Road, Pune - 411 014	*Rhizobium* *Azospirillum* *Azotobacter*, PSM
33	Madras Fertilizer Ltd. Manali, Chennai - 600 068	*Rhizobium*, *Azospirillum*, *Azotobacter*, PSM
34	Maharashtra Agro Industries Development Corp., Sugar Factory, Aarey Milk Colony, Dinkar Desai Marg, Goregaon, Bombay - 400 065	*Rhizobium*
35	Mahatma Phule Krishi Vidyapeeth Dept. of Agricultural Bacteriology College of Agriculture, Shivajinagar, Pune - 411 005	*Rhizobium* BGA

36	Micro Bac India, Shyam Nagar, 24, Parganas – 743 127, West Bengal	*Rhizobium* *Azotobacter*
37	MP State Agro Industries Corp. Ltd. Bio Fertilizer Plant, Indrapuri, Bhopal – 462 022, M.P.	*Rhizobium*
38	MP State Co-Op. Oilseed, Bio- Fertilizer Plant, Indrapuri, Bhopal – 462 022, MP	***Rhizobium***
39	Monarch Bio-Fertilizers and Research Centre, No. 12, SIDCO Ind. Estate, Thirumazhisai, Chennai – 602 107	*Rhizobium* *Azotobacter* *Azospirillum* Phosphobacteria
40	NAFED Bio Fertilizer, Plot No. 51-A, Sector 'F', Sanwer Road, Indore – 452 003, M.P.	*Rhizobium* *Azotobacter*, PSM
41	National Fertilizers Ltd. Vikas Tower, Indira Office Complex, Navlakha, Indore, M.P.	*Rhizobium* PSM
42	Nitrofix Laboratory, Block D, Jayshree Park, Calcutta – 700 034	*Rhizobium* *Azotobacter*
43	Orissa Agro Indistries Corp. Ltd., 95, Satya Marg, Bhubneswar – 751 007	*Rhizobium*
44	Cadila Corporate Campus Sarkhej-Dholka Road Bhat, Ahmedabad – 382 2210	VAM, PSM
45	Punjab Agro Industries Corp. Ltd. SCO 315-16, Sector 35-B, Chandigarh – 160 001	*Rhizobium*
46	Punjab Department of Agriculture C/o, Punjab Agricultural University Ludhiana – 141 004, Punjab	*Rhizobium*
47	Rajasthan Department of Agriculture Regional Research Station Durgapura, Jaipur, Rajasthan	*Rhizobium*
48	Rajasthan State Agro Ind. Corp. Ltd. 9-1, Subhash Nagar, Jotwara Road, Jaipur – 302 006, Rajasthan	*Rhizobium* *Azotobacter*

49	Rallis Agro Chemicals Research Station, 21-22, Peenya Ind. Estate II, Bangalore - 560 058	*Rhizobium* *Azospirillum* *Azotobacter*, PSM
50	Rashtriya Chemicals & Fertilizer Ltd., Chembur, Bombay - 400 071	*Rhizobium* *Azospirillum* *Azotobacter*, PSM
51	South Indian Mills Association (SIMA), Cotton Research and Development Coimbtore - 641 018, Tamil Nadu	*Rhizobium* *Azospirillum* *Azotobacter*, PSM
52	Southern Petrochemical Industries Corp., Ltd. 97, Mount Road, Guindy Chennai - 600 032, Tamil Nadu	*Rhizobium* *Azotobacter*
53	T. N. Department of Agriculture Bacterial Culture Production Unit Kudumiamalai, Tamil Nadu	*Rhizobium* *Azotobacter*
54	T. N. Department of Agriculture Bacterial Culture Production Unit Salem, Tamil Nadu	*Rhizobium* *Azotobacter*
55	T. N. Agriculture University Bacterial Culture Production Unit Coimbtore - 641 003, Tamil Nadu	*Rhizobium, Azolla* *Azotobacter* *Azospirillum*, BGA
56	T. Stanes & Co. Ltd. 8/23-24, Race Course Road, Coimbtore - 641 018, Tamil Nadu	***Rhizobium*** PSM
57	University of Agricultural Sciences Director of Research, G.K.V.K. Campus, Bangalore - 560 065	***Rhizobium***
58	University of agricultural Sciences Krishi Nagar, Bharward - 580 065	***Rhizobium***
59	U. P. Department of Agriculture Soil Testing Laboratory, Badaun - 243 001, U.P.	*Rhizobium*
60	U. P. Department of Agriculture Soil Testing Lab., Ramghat Road, Aligarh - 202 001, U.P.	*Rhizobium*

61	U. P. Department of Agriculture Regional Agricultural Testing, B-2, Adarsh Nagar, Meerut - 250 001, U.P.	***Rhizobium***
62	U. P. Department of Agriculture Soil Testing Lab, Mabadiya Nala, Sultanpur - 228 001, U.P.	*Rhizobium*
63	U. P. Department of Agriculture Regional Soil Testing Lab., Alambagh, Lucknow - 226 001, U.P.	*Rhizobium*
64	U. P. Department of Agriculture Regional Soil Testing Lab. and Demonstration Centre, Etawah - 206 001, U.P.	*Rhizobium*
65	U. P. Department of Agriculture Regional Soil Testing Laboratory, Civil Lines, Jhansi - 284 001, U.P.	***Rhizobium***
66	U. P. Department of Agriculture Regional Agricultural Testing Lab. and Demonstration Centre, Collectri Farm, Varanasi - 221 001, U.P.	*Rhizobium*
67	U. P. Department of Agriculture Regional Agricultural Testing and Demonstration Centre, Azamargh - 276 001, U.P.	*Rhizobium*
68	U. P. Department of Agriculture Soil Testing Laboratory, Civil Lines, Banda - 210 001, U.P.	***Rhizobium***
69	Vimpex Agencies, No. 3, Balaji Nilaya, 1st Cross, Swimming Road, Malleswaram, Bangalore - 560 003	*Rhizobium*, PSM *Azotobacter* *Azospirillum*
70	Rajashree Sugars & Chemicals, Coimbtore, Tamil Nadu	*Rhizobium*, PSM *Azotobacter* *Azospirillum*
71	Esvin Advanced Technologies Limited, Erode, Tamil Nadu	*Rhizobium*, PSM *Azotobacter* *Azospirillum*

72	K.C.P. Sugars & Industries Corpn. Ltd., Vuyyuru, Andhra Pradesh	*Rhizobium*, PSM *Azotobacter* *Azospirillum*
73	Krishna Industrial Corporation Limited, Nidadavole, Andhra Pradesh	*Rhizobium*, PSM *Azotobacter* *Azospirillum*
74	Sujay Bio - Tech Private Limited, Vijayawada, Andhra Pradesh	*Rhizobium*, PSM *Azotobacter* *Azospirillum*
75	Karnataka Compose Dev. Cor. Ltd., (K.C.D.C.) Bangalore, Karnataka	*Rhizobium*, PSM *Azotobacter* *Azospirillum*
76	Karnataka Agro Industries Corpn. Ltd., Bangalore, Karnataka	*Rhizobium*, PSM *Azotobacter* *Azospirillum*
77	Agro Bio Tech Private Limited, Kottayam, Kerala	*Rhizobium*, PSM *Azotobacter* *Azospirillum*
78	Narmada Bio-Tech, Inodre, Madhya Pradesh	*Rhizobium*, PSM *Azotobacter* *Azospirillum*
79	Madras Fertilizer Limited, Bangalore, Karnataka	*Rhizobium*, PSM *Azotobacter* *Azospirillum*
80	T. N. Department of Agriculture Bacterial Culture Production Unit Sakkottai, Tamil Nadu	*Rhizobium*, PSM *Azotobacter* *Azospirillum*
81	T. N. Department of Agriculture Bacterial Culture Production Unit Trichy, Tamil Nadu	*Rhizobium*, PSM *Azotobacter* *Azospirillum*
82	T. N. Department of Agriculture Bacterial Culture Production Unit Ramanathapuram, Tamil Nadu	*Rhizobium*, PSM *Azotobacter* *Azospirillum*
83	T. N. Department of Agriculture Bacterial Culture Production Unit Tamil Nadu	*Rhizobium*, PSM *Azotobacter* *Azospirillum*

Microbial Pesticides Producers in India

1. Anand Biocontrol Lab, Jalgaon
2. Ankur Seeds Pvt. Ltd., Nagpur
3. Agriland Biotech Ltd., Baroda
4. Bassaras Biocontrol Res., Pennailam
5. Bio control Res. Lab, Bangalore
6. Bio Era Technologies, Nagpur
7. Crop Health, Ghaziabad
8. Green Tech. Agro Products Coimbtore
9. Harit Biocontrol Lab, Yeotmal
10. Indore Biotech Pvt. Ltd., Indore
11. Innovative Pest control Lab, Bangalore
12. K.N.S. Biotech, Degloor
13. Kamul Herbal and Pheromone, Hyderabad
14. Kadur Agro Vision, Bangalore
15. Karnataka Biopesticides Ltd. , Hydrabad
16. Green Tech Agro Product, Coimbtore
17. Biotech International Ltd., New Delhi
18. Ganesh Biotech Control, Gondal
19. Krishna Biotech Ltd., Rajkot
20. Super Agro Safe India, Ahmendabad

—•—

Annexure III
(Application for the Registration of Microbial Pesticides)

Any person desiring to import or manufacture any insecticide may apply to the Registration Committee for the registration of such insecticide and there shall be separate application for each such insecticide. Provided that any person engaged in the business of import or manufacture of any insecticide immediately before the commencement of this section shall make an application to the Registration Committee within a period of [seventeen months] from the date of such commencement for the registration of any insecticide which he has been importing or manufacturing before that date. [Provided further that where any person referred to in the preceding proviso fails to make an application under that proviso within the period specified therein, he may make such application at any time thereafter on payment of a penalty of one hundred rupees for every month or part thereof after the expiry of such period for the registration of each such insecticide.] Insecticide Act, 1968.

On receipt of any such application the registration of an insecticide, the Committee may, after such inquiry as it deems fit and after satisfying itself that the insecticide to which the application relates conforms to the claims made by the importer or by the manufacturer, as the case may be, as regards [on such conditions as may be specified by it] and on payment of such fee as may be prescribed, the insecticide, allot a registration number thereto and issue a certificate of registration in token thereof within a period of twelve months from the date of receipt of the application. Provided that the Committee may, if it is unable within the said period to arrive at a decision on the basis of the materials placed before it, extend the period by a further period not exceeding six months. Provided further that if the Committee is of opinion that the precaution claimed by the application as being sufficient to ensure safety to human beings or animal are not such as can be easily observed or that notwithstanding the observance of such precautions the use of the insecticides involves serious risk to human beings or animals it may refuse to register the insecticide.

A. In the case of applications received by it prior to the 31st March, 1975 notwithstanding the expiry of the period specified in sub-section (3) for disposal of such applications, it shall be lawful and shall be deemed always to have been lawful for the registration Committee to dispose of such application at any time after such expiry but within a period of one year from the commencement of the Insecticides (Amendment) Act, 1977 (24 of 1977) Provided that nothing contained in this sub-section shall be deemed to make any contravention before the commencement of the Insecticides (Amendment) Act, 1977 (24 of 1977), of a condition of a certificate if registration granted before commencement, an offence punishable under this Act.

B. Where the Registration Committee is of opinion that the Insecticide is being introduced for the first time in India, it may, pending any inquiry, register it provisionally for a period of two years on such conditions as may be specified by it.

C. The registration Committee may, having regard to the efficacy of the insecticide and its safety to human beings and animals, vary the conditions subject to which a certificate a registration has been granted and may for that purpose require the certificate-holder by notice in writing to deliver up the certificate to it within such time as may be specified in the notice.

For more information on registration act and procedure visit site **http://cibrc.nic.in** *OR* **cibsecy@hub.nic.in**

Bioinsecticides Registered Under/Section 9 (3) of the Insecticide Act, 1968 As on 1st June 2005:

S. No.	Name of the Pesticide
13	*Bacillus thuringiensis* (*B.t.* & *B.s.*)
15	*Beauveria bassiana*
181	*Trichoderma viride*

Information on Registered Biopesticide and other insecticides in India please visit site **http://cibrc.nic.in/reg_products.htm**

Contacts for Registrations

Office Address	Secretary Central Insecticides Board & Registration Committee FARIDABAD, U.P., India	E-mail Address
Dr. Sandhya Kulshrestha	Secretary, CIB	sandhyak@hub.nic.in
Dr. D. Kanungo	Consultant	dkanungo@hub.nic.in
Dr. R.B. L. Bhaskar	Joint Director (Bio)	rbhaskar@hub.nic.in
Dr. Brajendra Singh	Entomologist (Insecticides)	brajsingh@hub.nic.in
Shri Vipin Bhatnagar	Joint Director (Chem)	vbhatnagar@hub.nic.in
Er. S.K. Ghosh	Deputy Director (Engg.)	skghosh@hub.nic.in
Dr. (Mrs.) G. Mukherjee	Deputy Director (Chem)	gmukherjee@hub.nic.in
Dr. B.S. Phogat	Deputy Director (Weed Science)	bsphogat@hub.nic.in
Mrs. Kanta Arora	Administrative Officer	karora@hub.nic.in
Shri Subhash Chand	PS to Secy. CIB&RC	schand@hub.nic.in
Shri Balbir Singh	Section Officer	bsingh@hub.nic.in
Shri Neeraj Kulshrestha	Legal Asstt.	nkulshrestha@hub.nic.in

For More Information : rvvyas@aau.in,rajababuvays@yahoo.com

RELATED SITES FOR INFORMATION

www.pesticideinfo.org
www.usda.gov
www.epa.gov

Note: For Biofertilizer registration information refer FCO Amendment order dated 24th March 2006 Schedule III (see clause 2(h) and (q) part A to D

—•—